AF505441

BOOK OF THE BUSH

Edward Kynaston is the author of the
prize-winning biography of
Ferdinand von Mueller, *A Man on Edge*.

BOOK OF THE BUSH

EDWARD KYNASTON

REED

This completely revised edition
Published 1990 by
Reed Books Pty Ltd
3/470 Sydney Road, Balgowlah NSW 2093

First published by Penguin in 1977

National Library of Australia
Cataloguing-in-Publication data

Kynaston, Edward.
(The Penguin book of the bush). Book of the
Bush

Includes Index.
ISBN 0 7301 0323 4
1. Australia – description and travel –
Guide books. I. Title. II. Title: Book of
the bush.

919.40463

Managing Editor Louise Egerton
Designed by Robert Taylor
Illustrations by Alistair Barnard
Typeset in New Zealand by Toptype,
Christchurch
Printed in Singapore
through Imago Productions (F.E.) Pte Ltd

CONTENTS

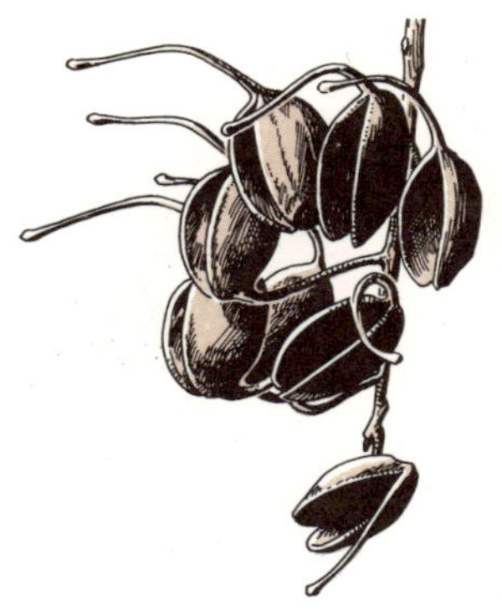

PREFACE

Australians are amongst the most urbanised, suburbanised, mechanised and dependent people in the world. We occupy one of the last great empty spaces of the world yet more than half our population is huddled in six great coastal cities. In most States the inland areas within easy reach of the capital cities all contain magnificent country, still largely in its natural, unspoiled, original form. Our climate is for the most part excellent and we spend a good deal of our free time out of doors. The sheer size of the interior is still quite frightening and perhaps plays a part in the extremes of gregariousness we sometimes display. This fear of the peace and quiet, the isolation and silence of the great free spaces shows itself both in a genuine respect for the pioneers of the pastoral past but also sometimes in an imitative travesty of their activities.

We have ignored or forgotten so much. We have forgotten where the sun rises and the moon sets, what weather the clouds suggest, and the compass corners from which the winds blow. Our ignorance of the daily commonplaces of natural living is demonstrated as a dangerous void when we try to answer what were once simple and ordinary questions. Which way does your home face? What is the prevailing wind in your part of the country? Those clouds passing across the sky; what sort are they? How far is it *on foot* to your closest friend's place, in time? In distance? When was the last full moon? In which direction does your bed lie? At what times will the sun rise and set today? In which direction would you look tonight to find the stars of the Southern Cross?

We have forgotten the simplest things about living comfortably and enjoyably at ease with our natural environment. In movement we have become almost totally dependent on machines: on buses, cars, trains, lifts, escalators, motorcycles, motor boats, on almost everything except our legs and feet. From home to work to entertainment and home again we carry with us, and are dependent on, an artificial environment. We live dangerously apart from the real life of the real preservation of nature. An enlightened and intelligent self-interest dictates this as a first necessity. In Australia, at least, it may still not be too late to halt the destruction of nature which in some other countries may already have assumed an irreversible trend. The first responsibility on all those who venture out amongst the hills and in the bush is to take thought for the future, their own and that of generations to come, by remembering and practising always the motto of the genuine conservationist and camper: *take nothing but photographs; leave nothing but footprints.*

Living simply, as part of nature where we belong, does not necessarily mean living simple-mindedly. To deliberately strip one's life to the essentials whenever this is possible is to put aside the plethora of objects that intervene between most of our daily reactions and a true awareness of ourselves. It is to replace the inauthentic by the authentic. It is to know the acutely rewarding feelings of satisfaction; aesthetic, practical, and philosophical, of extreme utility and simplicity. It is to know the land well enough to be able to live at ease with its dangers and its rewards while being acutely aware of the presence of both. It is a way of being fully alive, of forgetting totally the bizarre artificial preoccupations that we regard as normal in daily life, because we are living wholly and fully in the moment of *being-in-nature,* which will not in the end let us be anything but our true selves.

As with human beings, there is nothing more complex about nature than its simplicity. This is the paradox we need always to remember if we are to begin to attempt to live with, and understand, both nature and ourselves.

THE MAKING OF THE LAND

Far and few, far and few,
Are the lands where the Jumblies live;
Their heads are green, and their hands are
blue,
And they went to sea in a sieve.
EDWARD LEAR

One day, many years ago, while driving across the Nullarbor plain in the heat of midsummer I stopped my car to have a cup of tea by the side of the sandy road. When the dust had been washed out of my mouth I balanced the still half-full cup on the nearest mudguard and reached for some sandwiches. The cup slipped and fell and the tea turned the sand a darker shade of red. As I picked it up I saw amongst the red of the sand small flecks of white. I picked up a handful of wet sand to find it full of tiny, delicately perfect, white sea-shells.

The wind was scorching hot. All around was the flat dry aridity of the southern edge of the plain where fish had once flourished and swum and hunted in the cool dark depths of a sea. Here tiny shellfish had lived and moved and had their micro-milliseconds of life amongst waving weeds and fish forms long extinct, before falling to the sea bed to lie undisturbed forever.

It hardly seemed believable. There wasn't a tree in sight, only low saltbush through which the oven-hot wind was moving and sighing, and in the vast blue emptiness of the sky a wedge-tailed eagle slowly circling. Yet this had once been part of an area of Australia which had subsided beneath a sea full of minute marine organisms which lived and died and sank until their tiny bone bodies built up into beds of limestone, some over 300 metres thick. Where once water had slipped and glinted and sparkled coolly in the sunshine there was now only a flat plain carrying, amongst other things, the 530 kilometres of the longest straight stretch of railway track in the world, and a dryness that persists despite the small regular rainfall, which is quickly absorbed into the limestone beneath the sand.

The hot wind blew on that lonely road and on the sense I felt of the extreme ancientness and mystery of this solitary south land. What then seemed so immeasurably old turned out to be relatively young in the geological scheme of things. But Australia was incomparably older and more mysterious than even these tiny magical sea-shells, surfacing on the sandy edges of the road after so long, had first suggested.

The beginnings of Australia are as mysterious as those of all the other continents. Some time, some 6000 million years ago, the crust of the earth must have settled down from the continual rages and angry jerkings of infancy and settled into a period of relative somnolence and slow, deliberate movement. Then began the ripplings and wrigglings of the earth's surface that resulted in the first oceanic chasms and the first great chains of mountains. For the first time there was weather and weathering, water flowing to the oceans to become clouds and to be recycled again, causing erosion and the laying down of sediment.

This was the beginning of the Precambrian era that was to continue for over 5000 million years while Himalaya-size mountains rose and were worn down again (at least six times in this period), and climates changed from hot to cold and back again, and the seas moved uneasily about the everchanging chasms and depressions that each slow new upheaval created, and the great glaciers ground majestically back and forth a few centimetres, or a few metres, each year. It was towards the end of the Precambrian period, about 2500 million years ago, that life first appeared on earth.

About this time the beginnings of Australia appeared in the ancient plateau region of the Yilgarn shield of Western Australia. This was before there was any land mass remotely recognisable as the Australia we know today. The Yilgarn shield, which covers a good deal of the western part of Western Australia is one of the most ancient land surfaces in the world, having remained unsubmerged by the seas for most of this almost unimaginable span of time.

To look back so far is to look into an almost impenetrable darkness which can only be inadequately lighted by the flickering flares of the torches of conjecture. There was an area of the globe in which Australia would one day appear and the geological guess is that the area was mostly sea, studded with islands, and the Yilgarn shield at the edge of a great continent. The difficulty of speculation can be better understood by the realisation that the inclination of our planet towards the sun at that time is unknown, and it is perfectly possible that the Australian area was at one of the poles or on the equator or in the northern hemisphere. The one definite thing about which geologists seem to feel quite sure is that there was a continental mass reaching into the area and that the Yilgarn shield was part of it.

As time beyond human imagining passed, the slow changes continued. Mountain chains were thrust up and worn down, large masses of land appeared and disappeared again below the waves, volcanoes rumbled, poured out molten magma and convulsed and changed the landforms, spreading giant sheets of palely bubbled lava. In the seas the rain of fine particles never ceased, building up, grain by infinitesimal grain, sedimentary deposits up to 9000 metres thick! Later again the sea bed became land and the land was inundated yet again while the sleep of time passed so slowly that it hardly existed.

At least twice, while the land twisted and turned in its seemingly unending slumber, a great coldness covered the area and glaciers and ice-aprons covered the land, and huge icebergs drifted slowly through the waters. At about the time of this first glaciation it is believed there appeared a large continent, the first *Terra Australis*. Relentless, hardly noticeable, time broke it into fragments and scattered them through the seas until they formed the beginnings of the land mass that would eventually become the Australian continent we know today. This last event happened quite recently, only about 600 million years ago!

The long sleep of rock and air and water was coming to an end. Life, in its simplest water-borne forms, began to make its first appearance, and the land mass changed and kept on changing; sometimes split by a large sea, sometimes stretching far northward beyond where Papua New Guinea is today. The mountains rose and fell and in all the slow ceaseless activity of many millions of years coal and metals like molybdenum were laid down, together with the gold that was to help populate a quite different continent somewhere beyond the vast horizon of the time that kept on passing while climate changed and life went on evolving.

Then there came a time, about 200 million years ago, when great insects like *Mesotitan*, with a wingspan of over 30 centimetres, flickered through the steamy air of swamps and inland lakes, and about half of the area of present-day Australia was covered by a gigantic lake that we now call Lake Walloon. This lake is believed to have been one of the largest stretches of fresh water ever to have existed on the planet.

Soon, too, the giant dinosaurs reached the area. The bones of one, an herbivorous dinosaur, christened *Rhaetosaurus* were found at Durham Downs near Roma in Queensland. The bones weighed more than a tonne and a reconstruction showed *Rhaetosaurus* to have been about 12 metres in length overall, 4 metres high, with a tiny head and brain at the end of a long, thick neck. The brain was little larger than a tennis ball and was so inefficient that it had to be backed up by a second nerve centre in the pelvic area of the back legs in order to control the body's latter parts. The thick tail dragged on the ground but diminished to whip thinness at its end. Few traces have been found in Australia of the big carnivorous dinosaurs, and none of the enormous *Brontosaurus*, largest of the reptiles, which was 23 metres long and weighed anything from 40 tonnes upwards. Nor have any traces been found of the great flying reptiles such as the pterodactyls or pterosaurs, whose wingspan was sometimes as large as 7 metres. Beyond the Australian area the first true mammals and birds were appearing, small warm-blooded creatures better able to adjust to climatic changes than reptiles, who would eventually inherit the earth from their larger, apparently invulnerable, reptilean forbears.

While this was happening the area was changing yet again, the land rising, lakes and swamps disappearing, the great Lake Walloon shrinking almost into invisibility in a dry, flat, arid land. The relatively brief dryness was only a pause, an indrawn preparatory breath almost, for the shock of a flood beyond the wildest nightmare of the most competently nautical Noah. The flood was not local; in a greater or lesser degree it affected the whole earth so that even in the heart of Asia, the Gobi desert, the land was covered by a sea. Australia disappeared under the swirling waters, leaving only a few major islands, most notably in the south-east including Tasmania, in the north-west in the Kimberly-Central Australia area, and in eastern Queensland. The sea that had swallowed up most of the land was shallow, but it lasted between 10 and 20 million years.

The great reptiles still roamed the earth and the seas, including the biggest and most ferocious carnivore ever to have existed, *Tyrannosaurus*, a creature 15 metres long and 7 metres high. In the northern hemisphere, amongst the first mammals, there had already appeared the marsupials, some of which came to Australia, there to remain in peace, isolated from the rest of the world by the final geological events which contributed in the last 60 million years to the Australian continent as we know it today. In comparison with the previous 5000 million years everything now seems to happen with great speed; land alternating with sea, volcanoes raging, mountains rearing up and being inexorably worn down again.

Finally the earth's climate began to get colder. The rapid drop in temperature led into the Great Ice Age in which much of the northern hemisphere experienced arctic conditions, and ice sheets thousands of metres thick covered most of Europe and North America, with huge glaciers thrusting out into the low country from the high mountains. The area of the ice was about 10 million square kilometres and at its thickest the ice was 3 kilometres deep. The Great Ice Age lasted between 600 000 and a million years. It was divided up into four periods by false springs in which the temperature began to rise but soon quickly fell again. The last period of the Great Ice Age ended only about 10 000 years ago.

Australia appears to have hardly been affected until the last phase of the major glaciations of the north and only two small areas were affected by glaciers. These were in Tasmania and in the vicinity of Mount Kosciusko. In Tasmania an area of 5000 square kilometres was affected. A single ice-cap covered the north-western part of the central plateau, but there were many smaller ice-caps and valley glaciers. On the mainland, the Kosciusko area was glaciated for only 50 square kilometres, the ice forming chiefly on the south-eastern slopes of the higher hills and not more than 30 metres thick. Lake St

Clair, in Tasmania, lies in one of the many glacial valleys carved out at this time. Hitherto Tasmania had often been joined to the mainland. As the Great Ice Age came to an end and the glaciers and ice sheets melted, the seas rose steadily for about 75 metres and Tasmania became an island once again, while Torres Strait was created in the north, and the huge, complex harbour and waterways of the Sydney area came into existence.

The steadily increasing warmth of the climate since the last phase of the Great Ice Age has been important for inland Australia. The great rivers dried up, Lake Dieri slowly dried out, the great sandy deserts round the centre came into existence, while the abundant wildlife withdrew towards the coasts. Only the wind still works now in the hot aridity of the interior, sculpting sand dunes over much of the area, some over 30 metres high, and absorbing the sparse occasional rainfall as if it had never been.

Today the south-east of Australia contains the highest, but also the most recently made, land on the continent. The Australian alps, culminating at their highest point in Mount Kosciusko, are different from the other alpine ranges of the world because of an absence of the rugged peaks and precipitous 'faces' that accompany mountains formed by folding of the earth's crust. The great curve of the Australian alps swings from Mount Kosciusko in New South Wales round to Mount Feathertop and Mount Bogong in Victoria. This area is the highest remaining portion of the tableland that was thrust up from Cape York to Tasmania, at the time of the Kosciusko uplift, and which makes up the Great Divide of today.

The general character of the alps is relatively gentle and rounded, softened and restrained by a million years of erosion. Here there are few dramatic transitions, only high rounded hills with bare stony summits which slope quite gradually away from the highest points in all directions, to a confusion of tall-treed, densely bushed ridges and valleys. Erosion: the minute action of simple forces indifferent to time; snow and rain, frost and ice, the inexorable alternations of heat and cold, have smoothed and gentled away any really rough edges the high tableland may once have had. The last Great Ice Age played only a small part in a small area in the making of today's alpine scenery. Yet not far to the south, separated from the mainland by Bass Strait, is a high tableland of dramatic forms, operatic peaks, rugged valleys, and romantically beautiful lakes, which was, and still is, a similar sort of tableland.

Most of Tasmania is made up of a high central plateau that slopes gently from north to south. In the south-west corner lies one of the roughest areas of country anywhere in the world, a chaos of peaks and valleys, swamps and lakes, tangled bush and the notorious *Bauera* and horizontal scrub: a true wilderness. Most of the remainder of the plateau from the western tiers of the north to the estuary of the Derwent in the south, an area 1000 to 1300 metres in height, is a largely bare, rocky landscape glinting with thousands of tarns and lakes, many rivers, and rugged peaks rising high above the tumbled bleakness of the plateau.

The Tasmanian tableland was part of the upthrust of the land that took place at the time of the Kosciusko uplift. Its character is totally different from that of the high country of the mainland because of the Great Ice Age and the effects of the glaciation, which covered a large area to a great depth. It was ice — huge, slow, and ponderous — which scarred and sculpted, ripped and abraded the upland into the diverse and incomparably beautiful shapes it shows today. Frenchman's Gap, Cradle Mountain, Mounts Pelion and Ossa, Federation Peak and Mount Anne are only a few of the most famous peaks created by the action of the ice that once covered much of the island. Since only about 10 000 years have elapsed since the last ice age, erosion has only just begun to modify, so far almost imperceptibly, the Tasmanian landscape.

The planet turns, the clouds move, the rain and the snow fall, the wind blows, the earth's crust moves, the oceans lap with rough tongues at today's land. Change is ceaseless, as it has been for at least the past 6000 million years. Today's land is the surface of foldings and faultings, aeons of erosion, fire that sets rocks burning, cold that sets glaciers rumbling and gouging, the bones of uncountable millions of small marine creatures falling through the waters of the seas in a strange snow that continued for millions of years, and immense pressures almost beyond reckoning. The changes have not stopped. They are going on all day and every day. We simply do not live long enough to be able to see the process fully in practice. Geological knowledge has to be of the mind and the imagination, but practically useful for all that.

The surface of our earth is like a cake whose ingredients are being ceaselessly reconstituted. The mixing and the working and the moistening and the cooking have created pockets of especial richness throughout the layers of the cake, and continued mixing through the ages has spread particles of these metaphorical nuts and cherries and raisins throughout the crust in rich variety and abundance. Minerals, from radioactive ores to the dead inertness of lead; rare precious metals, from gold and silver to platinum; precious and semi-precious stones, from diamonds to agates, are all crazily mixed up in the always moving crust of the earth.

GOLD which more than any other single factor contributed to the populating of Australia in the foundation century from 1788–1888 was first found in the Fish River, New South Wales in 1823, but only in small quantity and it was not until 1851 that payable gold was discovered and the rush began. Between 1851 and 1861 the population increased almost threefold. By 1877 gold had been found in Victoria, Tasmania, Queensland and Western Australia.

Gold is ancient as well as precious. The soft yellow metal almost twenty times as heavy as water was formed and deposited in the earth's crust anything from 50 million to more than 2000 million years ago. It is almost always found in rocks containing quartz veins and in veins of granite and gneiss, integral with the surrounding rock, and it has to be separated by crushing.

The first gold found in Australia was mostly alluvial gold in the beds of streams and rivers. Some gold that had reached the surface of the soil through movements of the crust was gradually broken down through weathering and washed into rivers where it was pushed along by the force of the water until it stuck fast on a clay or granite bottom, or sank amongst sand and gravel above rocky bars across the creeks or rivers. Gold is also deposited by water currents on the insides of curves, and where tributaries flow into a main stream.

Since this process has been going on for many millions of years, gold is also found in 'deep leads' which are riverbeds of past eras, now sunk deep into the earth, or if the crustal movements have been vigorous enough, thrust high up inside hills or mountains. Volcanic activity, with great lava flows, engulfed many of the rivers of the past.

Alluvial gold is the chief interest of the amateur prospector since it is easy to pan for, and panning can be carried out effectively using a simple metal or plastic soup plate or shallow dish.

PRECIOUS AND SEMI-PRECIOUS STONES: sapphires, garnets, zircons, diamonds, emeralds, tourmalines, chrysoprase, topaz, opals and many others have become the province of specialist amateurs banded together in gemhound clubs who seek their material in old quarries, spoil dumps, old mine workings, riverbeds, pits and beaches. Gemhunters carry a variety of equipment, including prospectors' hammers, sledge hammers, cold chisels, crowbars, shovels, sieves, miners' panning dishes, jars and phials, and magnifying glasses.

Gemstones are found all over Australia and because of the huge area and small population it is probably the best country in the world for the amateur gemhunter. Australian opal, in many varieties, is famous. There are large sapphire fields in eastern Australia, and a small area of diamonds in the Inverell region. In semi-precious stones, the agate deposits and the chrysoprase and jasper of Queensland are well known to collectors in many countries.

Most precious and semi-precious stones originate as a result of the effects of intense heat, pressure, contamination with other chemicals, and crystalline growth over huge periods of time. As an example, agate is formed as a deposit in cavities in volcanic rock. It is made up largely of silica which cannot be dissolved naturally. In volcanic conditions when there is immense heat and pressure the silica is dissolved and carried to the cavities in cooling lava. Here it is coloured by whatever metallic substances are also in solution. Agates are made up of minute fibrous crystals in layers of as many as 10 000 to the centimetre. Time is the perfecting factor in the making of many crystalline gems and semi-precious stones.

Everyone who looks for gemstones or gold *must* have a 'miner's right', to enable him lawfully to do so. This applies to all States. The 'miner's right' is a document issued on payment of a small fee to a State Mines Department, and is of historical and economic interest rather than legal significance, since it is vague in application, negative in its promises, and confers no substantial rights on the holder.

Time beyond our comprehending made the minerals, metals, and precious and semi-precious stones that are still to be found in Australia, as it made the hills and the valleys and the deserts, and the coastal plains where so many of us live today. Time separated Australia from a once great land mass; time isolated our continent in the south. The story of Australia is still largely the story of time made manifest in matter; time frozen in rocks, soil, and sand which are apparently static, changeless, timeless but actually dynamic, moving, time-driven in frantic, ceaseless, secret movement.

Further Reading

Charles F. Laseron, *Ancient Australia*, Angus and Robertson, 1969.
Charles F. Laseron, *The Face of Australia*, Angus and Robertson, 1969.
N. and A. Learmonth, *Regional Landscapes of Australia*, Angus and Robertson, 1971.
Brown, Campbell and Crook, *The Geological Evolution of Australia and New Zealand*, Pergamon Press, 1968.
H.K. Garland, *All About Prospecting*, Reed, 1982.
H. Neese, *Gold Fossicking in Australia*, Angus and Robertson, 1984.
R. Shears, *Gold: Where to Go and Where to Find It*, MacMillan, 1980.
J. Toghill, *Australia's Hidden Treasure*, Angus and Robertson, 1988.

PERSONAL EQUIPMENT

Man is the limiting factor.
UNKNOWN SCIENTIST

If you were going into the bush for a day, or a week, or a year for that matter, who would you choose to be your companion? What sort of attributes would you look for in the person of your choosing? If, for a longer or shorter time, you are going to be stuck with someone from whom you can't get away very easily then obviously the choice is going to be important, if not crucial, to your enjoyment and possibly even to your safety and well-being.

The reactions and behaviour of people in town and in the bush are often radically different. Therefore the choice of companion is not only an important one, but often difficult as well.

The most important person you ever accompany anywhere is yourself. Are *you* the sort of person you would want to go into the bush with for any length of time? When all the familiar things are gone and the big quiet stillness of the bush surrounds you, how will you react? Will you want to raise your voice and talk loudly to try to fill the disquieting and unfamiliar silence? Some people do. Are you likely to get irritable and bad tempered and jumpy, without being properly aware of it, because familiar comforts are absent or the physical effort you are making is strange and difficult? Some people do. Will you become domineering and dogmatic and loudly insuffer-able because the bush makes you feel insecure and afraid and you can't admit the *natural* fear of the unknown? Some people do.

P ERSONAL EQUIPMENT is vitally important in servicing the body, but along with the body we take the often neglected or taken-for-granted imagination, feelings and other attributes of personality. Newcomers to the bush bring with them often, as their first and most intimate companions, fear, insecurity, ignorance, thoughtlessness and inexperience. There isn't much that can be done about the inexperience initially, except to be fully aware of it, but the other difficulties can all be dealt with successfully through intelligent effort and awareness of ourselves, and by a respectful familiarity with the land and what it contains.

In order to begin pioneering anew, the first thing that has to be done is to make sure that one's most personal equipment is in order. Personal stocktaking of attitudes and ideas, as well as prejudices and probable reactions, is as important as the basic preparation for a journey of discovery that at some time or another is bound to lead to unexpected self-revelation. Journeys into the bush can also be journeys into our inmost being. The most unexpected thing we are likely to find in nature is ourselves. It matters very much who you are, if you don't already know who you are.

Personal equipment is meant to serve and comfort you. It is not taken along to impress others, feed vanity, or bring the luxuries of suburbia to the bush. The most important bit of personal equipment you can have is what goes on your feet. Feet are the only natural form of transport in the bush. To get out of cities and towns it may be necessary to use aircraft or trains or cars, but once in the open, feet come into their own, and what you choose to wear on them for support, protection and sure-footedness matters immensely.

Comfort in every conceivable condition is the first need in footwear. Ideally you should be able to put on footgear and not think about it until you take it off again. The next requirement is firm support over, and grip on, all sorts of yielding and unyielding surfaces, from mud to rock, from sand to weedy creek stones and pebbles in, and out of, rushing water. Equally important is the protection and support given to feet and ankles against rocks, fallen branches, undergrowth, and the rugosities that occur whenever there is hard going. The right footgear makes the going easy and journeys seem short. The wrong footgear makes the going hell, short journeys long, and saps energy away in pain. The right footgear is always worth about ten times what you pay for it.

BOOTS in general cannot be beaten for serious walking. Only actual physical defect should lead to other footgear. Unfortunately, having arrived at the conclusion that boots are the only possible choice, we have also landed in an area of variety, confusion and even chaos. Boots come in every imaginable shape, leather, design, and lacing. Some come padded, some come unpadded, some have uppers reaching high up the calf, some have uppers that barely reach the ankle, some have loose tongues, some have sewn-in tongues; there are also double tongues, and split tongues; some have lacing by eyelets, some have lacing via rings, hooks, and clips, some have several layers of leather in the sole, some have only one, and a thin one at that.

What the boot is to be most commonly used for is the consideration that cuts through all the confusion. So far as this book is concerned there are three sorts of boots: for walking, scrambling, and climbing. Since climbing is a mystic and highly specialised discipline all its own, only boots for walking and scrambling are left. What is required is a boot that is adequate for both purposes. Beyond this, Australian conditions make difficult demands. For much of the year light boots suitable for dry and warm to hot conditions are needed. For the remainder of the year, and more in some years than others, a heavier and more robustly designed and constructed boot is needed for cold and wet conditions. Ideally an all-round walker should have both. In fact what happens, particularly amongst younger walkers, is that a heavy, cumbersome, unnecessarily complicated boot, usually soled with what feels like lead, is chosen, in an attempt to meet every possible condition from tropics to snow, from drought to deluge, from storm to sunshine. Other factors are at work here too, vanity in possessing such mightily impressive looking boots, reassurance because the boots give such a solid ground-shaking impact as they descend and meet the earth, and the aura of seasoned toughness that is presumed to radiate from them on to their owner.

Unquestionably this is putting the boot on the wrong foot. If limited to one pair of boots, the first thing is that they should be the *lightest* possible to meet all conditions. My personal preference is for a pair of boots that may be a little too light for really bad conditions which, unless you are a masochist, means for only a very small percentage of the time you actually spend out walking. In these boots you can usually get away with the odd occasion of extreme weather and heavy terrain, while for the rest (the majority) of your walking time you have the advantage and pleasure of a reasonably light boot.

During the war the synthetic rubber Vibram sole was introduced for mountain troops, and this was available commercially soon after the war, when it swiftly began to displace nailed boots from the scene. Since then, Vibram and Vibram-type soles have taken over almost completely. These soles are moulded to imitate a mixture of clinker and mugger nails and are thus primarily intended for use in climbing. Like most soundly based fashions they have spread themselves into areas for which they were never intended, so that today you may see giant-booted Vibram-soled walkers on sedate national park trails, on gentle bush tracks, and even rambling with groups along the roads round beauty spots (with trousers tucked carefully into socks to keep out scree and other gunk!).

General purpose nylon and leather lightweight walking boot.

Vibrams were a valuable innovation. Whether they suit the sort of *walking* you are liable to do is another matter altogether. Now here's some heresy. There are other sorts of solings (and much lighter ones) than the Vibram-type, many of which, in my opinion, are adequate and more than adequate for most of the generally warm and dry conditions of Australia, and which will do at a pinch, and briefly, for cold and wet conditions.

There are numbers of bushwalkers who now swear by light-weight walking or jogging shoes. These are wonderfully comfortable to walk in and some have recently become available in bootee form, which greatly improves their usefulness for bushwalking. In dry settled conditions over moderate terrain these shoes answer well, but in wet or rocky or very rough going they are bound to be inadequate.

Probably the best compromise for most sorts of going is the synthetic and leather lightweight walking boot (illustrated) which is comfortable, easy to break-in, rugged and dependable. The life of this boot is, inevitably, shorter than that of a heavier leather boot, but when relative cost is taken into consideration this is not a serious drawback.

Also available in Australia, and made here, is a desert-type boot with a low upper, a flexible sole that has a good, moderately deep, grip pattern, three layers of leather between bottom of boot and composition sole (good shock absorption when the foot is hitting the ground many thousands of times a day), and which is made of whole leather with a smooth tan. It is relatively inexpensive, can take occasional bad conditions (if treated properly), is very light — just over half a kilogram each boot in an average size — and is soundly made.

For regular, heavy going on high ground in cold, wet winter obviously a heavier boot with a sewn-in tongue, some padding for extra warmth and comfort (though combinations of different thicknesses of socks can serve almost as well), and an orthodox Vibram-type sole is what is needed. Boots of this sort are moderately costly, distinctly heavy (1.5 to 2.3 kilograms), but very durable. With proper care they will last for a good many years and remain good enough to enable you to tackle almost any terrain or conditions.

B UYING BOOTS can be a traumatic experience unless basic needs are kept in the forefront of the mind and repeated frequently to oneself. Basic needs are *comfort, light weight, sound construction.*

When trying boots make sure that they are long enough to allow for toes to slide forward, as often happens when going downhill, without coming into contact with the inside of the toecap. There are two ways of ensuring this. Lace up the boot tightly, stand on a sloping surface (the face of the box stool in the shop will do) and try to push your toes to touch the front of the boot. Alternatively, leave the boot unlaced, push the toes forward to the front of the boot and slide a finger down behind the heel. If the finger fits quite loosely the fit should be about right. If the finger is tightly gripped the fit will be too small. If two fingers, side-by-side, can be inserted, the fit will be too large.

The width of a boot can only be judged by feel, and should be quite firm in its fit across the foot but allow toes to be moved freely. Boots widen slightly with wear so a firm fit at the purchasing stage is essential. A good boot has plenty of toe room, comfortable length, a width that grips the sole of the foot firmly, but allows the toes to move freely. Above all a good boot feels comfortable and *right* on the foot. The purchase of boots which do not feel comfortable and right, about which you have the faintest suggestion of doubt, is a waste of time and money that will be regretted later. If in doubt keep trying on more boots!

Wear at least one pair of thick socks (they can usually be borrowed from the management) when trying on boots. If you are likely to regularly wear two pairs (one thin and one thick pair, or two thick pairs or some other combination), then wear what you will wear in the field. What is suitable for one person is not suitable for another, whatever the 'expert's' sock formula may be. Just remember that boots do stretch with use and try to make allowances for this when buying.

If, in spite of every care in their selection, your new boots refuse to behave themselves, you can always try the water torture on them. This involves filling them with *hot water* at about 60°C for about half a minute or a minute, then putting them on and walking in them for forty-five minutes. You can also soak them in *warm* water for half an hour and then wear them through the day until dry. The idea is to create conditions similar to those experienced on a long walk which help to mould the boots to the individual feet. This doesn't do much harm to the boots and may help, but if this sort of treatment is needed it suggests that the boots were probably wrong for you in the first place. Don't try any other 'breaking in' treatment. Greasing boots and baking them in a slow oven won't help. Soaking in alcohol will ruin them. Hammering hardnesses hopefully with a mallet or the nearest house brick will only relieve your feelings while leaving your boots and blisters as they were. If any of these latter alternatives begin to look attractive to you, give in gracefully: either unload them on your worst enemy, or bury them deep in the backyard, and begin all over again.

Boots are a big basic investment in both time and money. Your best insurance for them is care. Leather, like wood, is organic, responsive material which reacts well to bad treatment, but superbly rewards only average good treatment. Leather can, and must, be fed if it is to retain the suppleness and strength that allow it to mould itself to your feet and give the support, protection and insulation that is expected of it. Leather is a skin, the skin of *your* feet when made into boots and shoes, and like any other skin it has to be kept clean first of all and then kept well supplied with natural oils.

Boots should not be allowed to dry out with mud and dirt still on them. This is bad for the leather, as well as for any stitching and cementing. Brush or scrape off mud if possible before returning home. At home wash with cold water and cloth or brush as needed, and allow to dry in a place of moderate (not hot) temperature.

The most important part of boot care is in ensuring that the leather is never allowed to dry out completely. Dry leather is brittle leather which absorbs water like a sponge and quickly produces savage blisters. Leather left dry leads to the inevitable disintegration of the boot. Most boots need frequent applications of oils to keep them in their best condition, and most boots will respond to natural oils. Suede is an exception: it can only be treated with silicone compound water repellent. The same treatment is recommended by some people for chrome, or dry-tanned leather.

PUTTING A SOCK IN IT is the next stage after getting boots. Fortunately there are now huge quantities of socks being made out of every conceivable natural and artificial material and in mixtures of the two. There are also socks in thicknesses varying from the see-through to densely matted hairy knits for cold-blooded Eskimoes. The important thing is that it is now easy to get socks of any length, knit, material and thickness. The only answer to the sock problem is to experiment until you find the formula that suits you.

FOOT CARE is largely commonsense. Keep toenails short, keep feet clean, *dry* your feet, and dry between each toe after showering. When walking any distance use rest periods to air your feet or soak them in any cold water that may be available. In summer heat some people like to dust their feet with talcum or medicated powder. This doesn't suit everyone but it's worth experimenting to find out whether it suits you. When walking, prevention is the better part of commonsense. If trouble develops on any part of the feet, stop and find out what it is. Another sock may be needed, or a deep wrinkle taken out of an existing sock. There may be a bit of twig or small stone working its way around. It may simply be that the laces need tightening. Some blistering can be expected before feet toughen up, but whenever blisters occur *do not neglect them*. Use Band-Aids or cotton wool and sticking plaster to cover, protect, and insulate the area. *Don't plod on* because of what your companions may think, or because you don't want to delay them. If they are the right companions they won't mind. If they are the wrong companions it doesn't matter anyway. And if you try to go on you may end up delaying them longer than you imagined. Getting major medical help into the bush takes time, and evacuating a casualty takes even more time. Looking after yourself properly is your first move in looking after others. If you're camping you will obviously have spare dry socks and lightweight footgear to wear at the end of each day.

GAITERS can be useful in snow, when travelling through wet grass, working on small scree, or when wading streams. Gaiters can never be knee-length boots, but a short gaiter up to about 25 centimetres in length does give some protection and puts off the evil hour when moisture makes its way down socks and into boots. The most efficient aspect of gaiters is in keeping foreign bodies out of boots and keeping the feet and ankles clear so that they do not catch on obstructions.

C LOTHING is mostly a personal matter but also depends on the time of the year, the distance to be covered, the time spent away from home, the terrain likely to be encountered, the usual weather on the way. Comfort comes first, as it does with most personal equipment, and then efficiency. Basic clothing must be worn or carried. Basic clothing is clothing that will comfortably protect you from the worst that the local climate in the area you are entering can do. Excessive heat, cold, and wet are the enemies. These elements dictate basic clothing.

A pair of light shorts, or lightweight cotton trousers (sunburn is painful), and a light cotton shirt and some sort of head covering will meet hot conditions best.

Cold is best countered by woollen shirts, *woollen* pullovers in light weights and lightweight, windproof outer clothing.

Wet is best met by waterproof, windproof, outer clothing, which today means a parka and overtrousers of Gore-Tex fabric, or the cheaper, and it must be admitted, much less efficient oil-proofed japara.

If walking in Australia anywhere above about 1300 metres in summer, all the basic clothing should be carried: the cool clothing for wear during the heat of the day; the warm clothing for the evenings and nights; and the windproof and waterproof clothing for the sudden storms and changes that can occur anywhere in the mountains.

Of all the materials available, wool is the most valuable because of its warmth for weight ratio, its resistance to rain, and its retention of warmth even when soaking wet. Synthetic materials can be dangerously chilling in the wet, and in snow can contribute to frostbite. Cotton is comfortable for underwear and T-shirts or undershirts, and can make tolerable the ticklishness some people feel in direct contact with wool. Layers of thin clothing (cotton or wool or both) conserve heat best since the effect is to make several thin layers of insulation which trap dead air between them. Layering gives the most warmth for the least weight.

K EEPING YOUR HEAD AT A MODERATE TEMPERATURE in most conditions is important because it has a big influence on the heating of the rest of your body. The old saying that if you want to keep your feet warm you have to wear a hat has a good deal of truth in it. The head and neck together account for one-fifth of the body's blood supply and a quarter of the body's oxygen. Radiation (heat loss) from the area is substantial. If the head is cold the whole body will be very cold. In excessive heat, a cotton or towelling hat kept moistened with cold water which will be constantly evaporating, will cool the most critical area of the body. In extreme cold and wet, a woollen cap or knitted woollen balaclava covered by parka hood will maintain a comfortable head temperature and contribute largely to the heating of the rest of the body. In between, almost anything goes. A hat, or head covering of some sort is basic clothing, and should always be carried on anything more than an afternoon stroll.

UNDERWEAR should be comfortable and unrestrictive. If cold conditions are likely wool is best; in heat, cotton. In extreme cold, string underwear is particularly good but rough to wear against the skin. The string mesh traps a lot of air which makes a good insulating layer.

THE WOOLLEN PULLOVER is probably the most useful and versatile garment available to the walker, especially in regulating body temperature in the course of a normal walking day, when heat and chill accompany walking and resting, and as the temperature first rises and then falls. Two or three pullovers slipped on and off as required will usually make adjustment comfortable and easy.

Modern Gor-Tex parkas, wind and rain proof, with wired visor hoods.

T HE PARKA is the ultimate outer garment and some care is needed in its selection. Unquestionably, the best available today are the ones made from Gore-Tex material, a high-tech fabric. This fabric is a laminate of protective material with Gore-Tex in the middle. The Gore-Tex contains billions of microscopic holes, randomly offset, which are too small to allow rain to penetrate but large enough for perspiration vapour to escape. Most condensation, the curse of other materials, is avoided. The material is also almost completely windproof.

The design of the latest Gore-Tex parkas is generally first rate. They have large pockets, waist draw cords, good zipper and velcro closures, and excellent capacious hoods with wired visors which may be rigged to protect the face from the weather. Overtrousers of the same material and of the same first-rate design are also available.

For extreme conditions of cold, snow and ice, jackets padded with superdown with a Gore-Tex outer shell are available.

Oil-proofed japara parkas, for many years the mainstay of Australian bushwalkers, are still widely worn. They are relatively cheap, but inefficient when compared with their high-tech Gore-Tex cousins. Overtrousers are available in the same material.

When purchasing a parka it is advisable to make sure that the hood is ample enough not to exert any pull on the head, and to have the parka slightly oversize to allow for easy movement, and for the addition of clothes beneath if needed. Avoid rubberised or synthetic fabrics as they cause heavy condensation inside. Rubberised cloth is also heavy and hot, and plastic tears easily in rough conditions.

When buying overtrousers make sure that the legs are big enough to pull on over boots.

Parka and overtrousers are the best defence against hypothermia, or the rapid loss of heat which can lead swiftly to death from exposure. Hypothermia is a real danger in most of south-eastern Australia and for a longer period of each year than is generally realised. Cold by itself is unpleasant but not especially dangerous. It is when cold is combined with wind and wetness that the catastrophic heat loss of hypothermia occurs. Parka and overtrousers are the first line of defence against this condition and are not clothing upon which economies can be made without the possibility of endangering one's life.

P ERSONAL EQUIPMENT is the most important of all your equipment. Whether or not you feel comfortable and secure with yourself and with your equipment can make or mar any pioneering you intend to do. Equipment can be reduced or excluded quite easily if necessary, but you cannot leave yourself at home. If you are reasonably relaxed and at ease with yourself and what you wear and carry with you, your chances of successfully exploring both the terrain you choose, and yourself, are enormously improved.

The best guide to gear in Australia is *Equipment For Bushwalking and Mountaineering,* Melbourne University Mountaineering Club, 1982.
Jeff Carter's Guide to the Outdoors, Rigby, 1981.
R. Graves, *Australian Bushcraft,* Dymocks, 1984.
Paddy Pallin's Bushwalking and Camping, Sydney, 1985.
J.F. Meir, Wm. C. Brown, *Backpacking,* Iowa, 1983.

CHAPTER 3

THE MAN-MADE LAND

*And all lying mysteriously within the
Australian underdark, that peculiar lost
weary aloofness of Australia.*

D. H. LAWRENCE

South-eastern Australia is only 20 per cent of the surface of the
continent yet it contains 80 per cent of the population. On a midwinter morning in a small area of that 20 per cent you might find
deep snow round the icicle-hung chalets of the upper alps, an empty
blue sky, and freezing winds. Only 80 kilometres away and 2000
metres lower down a farmer is driving his tractor on pastureland
with his sleeves rolled up. A hundred or so kilometres further on a
brown hawk is hunting in warm air over the saltbush plains of the
semi-arid Big Desert, watching for movement of mouse or lizard or
large insect.

The distance travelled from alps to desert is not much over 500
kilometres yet in this relatively short traverse, landscape and climate
vary remarkably in as sharp contrasts as are to be found anywhere in
the world in such a limited area. The majority of the 16 million
people who live in the area live in a few large cities and towns and
mostly on the coast. Comparatively, the land behind the east coast is
empty of people, yet the few who have occupied the land have had
an effect out of all proportion to their small numbers.

The chain of hills and mountains running down the eastern side
of Australia is called the Great Dividing Range, or simply The Divide.
Time and weathering have reduced what was once a chain of some
of the largest mountains in the world almost into insignificance.
There is little land left over 600 metres in height now, and in some
places The Divide is only a vestigial remnant of its once lofty self.
Any lengthy journeys undertaken in Victoria or New South Wales
involve crossings, sometimes frequent, of The Divide somewhere
along its length. In Queensland, where distances are greater, The
Divide must still be crossed, if less frequently. Few Australians realise
that The Divide begins in the tropics at Cape York, extends right
down through three mainland States, and ends in a fourth State,
Tasmania. From the hot mud of the mangrove swamps of Cape York
to the windy top of the winter snows of Mount Wellington there are
3000 kilometres of climates of all sorts, a temperature range that is
staggering, and a great variety of landscapes.

The two dominating facts of the inland life of the east coast are
the highlands, the hills and mountains of The Divide, and the ever

The Great Dividing Range

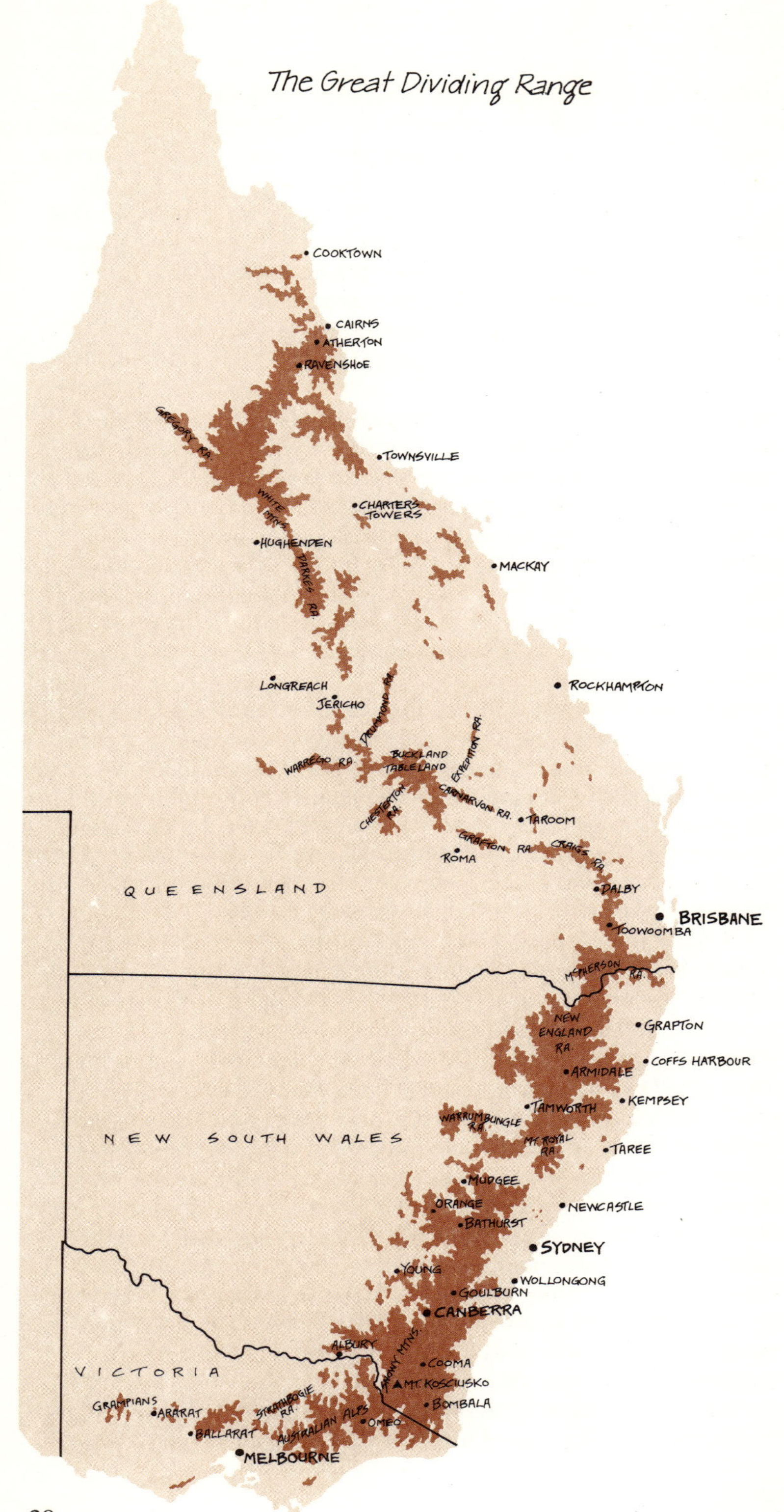

present bush. From tropical rainforests of great density and richness the bush ranges in all its varieties through subtropical rainforest to temperate rainforest, to both wet and dry sclerophyll forest (much of which has been cleared for farming) to the notorious horizontal scrub and *Bauera* of the south-west of Tasmania. Despite its seeming abundance, forest covers only about 4 to 5 per cent of the surface of Australia, making it a relatively disadvantaged country for which conservation is essential for adequate survival.

To the east of The Divide is the narrow strip of coast along which the majority of Australians live. To the west of The Divide the land slopes away in value and utility to Mallee and Mulga and other scrubland, and then actual desert regions: the Simpson, The Great Victoria, The Great Sandy Desert, the gibber-strewn horror of Sturt's Stony Desert; arid scourges of a generation of explorers.

Australia is the driest continent and its small rainfall is the least reliable on the face of the planet. Drought, prolonged and only statistically predictable, is the curse laid on its farming community. The east coast and The Divide are the areas most favoured by rain but even this is very variable and drought, and monsoonal storms slashing down from the tropics, alternate with provocative irregularity; the semi-desert of one year often becoming the floodland of another year.

THE FIRST PEOPLE. Some 50 000 years ago the first people appeared in Australia. They confronted the land. They entered into an abiding and rewarding relationship with it that was to last ten times as long as most known civilisations. Of all the pioneers they have been the most successful and the wisest. Lacking sophisticated technology through which, by force, they might have attempted to dominate the land, they instead used intelligence and wisdom. Without the noise of machinery in their ears, the Aborigines were able to hear the voice of the land and follow its laws.

The three primary needs of life were then, as now, in order: water, food, shelter. Each need-problem acted and reacted on the others creating complexities which laid down the patterns and possibilities of living. Water was a fundamental. Human beings can live for ten days without food, but only three without water. Water determined the areas in which game could be hunted, determined the movement of those areas as climate varied, kept the game moving, kept the men moving, shaped the portable shelters, created light, multi-purpose equipment. In the Aboriginal beginning there was movement, as there still is today. Even beside the abundance of the tidally fluctuating fields of the sea Aboriginal camps were kept moving, if only a few hundred metres at a time.

Despite much of what has been written about Aborigines in the past it now seems that they could have had settlements, and growing from them inevitably, the sort of civilisation that evolved elsewhere.

It seems clear that while there was not an abundance of food there was sufficient to make agriculture seem unwarranted! The Aboriginals *chose* mobility and a way of living at variance with most of the rest of humanity, or so it appears. It is also possible that movement had become as natural to them as breathing and that they conceived of year-round immobility as death. Whatever the reason, they learned to live with the land, conjuring water from hidden sources in plants and trees, gathering plants, nuts, leaves, berries; catching small rodents, snakes and lizards; hunting kangaroo and wallaby and the larger creatures, but always, for all the thousands of years of their tribal existence, moving, moving, moving, until movement itself became the one unvarying permanence and security of their lives.

There were pockets of semi-continuous settlement, but these were very few. One, to which we owe most of our recently acquired knowledge of the earliest Aboriginals, occurred at Lake Mungo, north-east of Mildura in Victoria. Recent archaeological finds at Lake Mungo place the Aborigines as the first men to practise cremation, as the makers of the first ceramics, as the makers of the first edge-ground stone axes, as the creators of some of the earliest rock carvings, and as the first men to utilise ochre for their art. The Lake Mungo people seem to have followed a seasonal movement between fishing in the lake in summer and hunting and collecting eggs on the plains in winter.

It can hardly be doubted that social and religious beliefs developed early (the cremation evidence at Lake Mungo suggests strongly the existence of beliefs about life and death) and persisted for immensely long periods of time closely interwoven in the complicated ballet-ritual the Aborigines created as they danced the seasons round, weaving subtly complex, repetitive patterns of movement across their tribal areas, in the eternal now of the land's seemingly unlimited time and space. The Aborigines seem to have been more than intelligent enough to envisage change but happy enough in the satisfactions of their harmonious relationship with the land not to desire it.

L IVING WITH THE LAND. The first Australians lived unobtrusively in a creative relationship with nature. In all the many thousands of years of Aboriginal occupation there occurred only the natural changes brought about by time, tide, the weather, and the movement of the earth itself. The only disturbances that the Aborigines are known to have brought to the ecology were the dingo, and burning off. The dingo was intended to be an aid to the hunters, but was more commonly a simple domestic pet. The dingo has been blamed for the destruction of some native fauna, but there is little evidence to support this. Burning off was a practice carried on by Aborigines since earliest times and was best described in its

practice and consequences by white explorer-pioneer Major
Mitchell in 1848.

> *Fire, grass, kangaroos, and human inhabitants,
> seem all dependent on each other for existence in
> Australia; for any one of these being wanting, the
> others could not longer continue. Fire is necessary to
> burn the grass, and from those open forests, in
> which we find the large forest-kangaroo: the native
> applies that fire to the grass at certain seasons, in
> order that a young green crop may subsequently
> spring up, and so attract and enable him to kill or
> take the kangaroo with nets. In summer, the
> burning of long grass also discloses vermin birds'
> nests &c., on which the females and children, who
> chiefly burn the grass, feed. But for this simple
> process, the Australian woods had probably
> contained as thick a jungle as those of New Zealand
> or America, instead of the open forest in which the
> white men now find grass for their cattle, to the
> exclusion of the kangaroo, which is well-known to
> forsake all those parts of the colony where cattle
> run. The intrusion therefore of cattle is by itself
> sufficient to produce the extirpation of the native
> race, by limiting their means of existence; and this
> must work such extensive changes in Australia as
> never entered into contemplation of the local
> authorities.*

NEW INVADERS. The dreaming and dreamtime of the open
forests, for so long maintained by the Aborigines, were coming
to an end, and the convulsion of relatively rapid change was about
to alter, with unprecedented speed, the face of the land for ever. The
instinctive conservationism of the Aborigines was about to give way
to the compulsive, brutally exploitative 'rational' agriculture of
people from a crucially different climate, landscape and culture.
Small wind-driven wooden ships, crewed by determined adventurers
and using compass and cross-staff for their primitive navigation,
moved out from ports on the other side of the northern portion of
the planet and across great waters hitherto largely unexplored.
Global space was in the process of being conquered, but time was
still the enemy of the navigator. The chronometer, from which
longitudinal position could be precisely established, had still to be
invented and perfected before the conquests of terrestrial space could
be consolidated by colonisation. The Aboriginal dreamtime was soon
to be turned into unimaginable nightmare.

Describing the Cape York area in 1623 one early explorer, Carstensz, a Dutchman, wrote that the land that he saw was:

*a barren and arid tract, without any fruit trees
and producing nothing fit for the use of man; it is low-
lying and flat, without hills or mountains, in many
places overgrown with brushwood and stunted wild
trees; it has not much fresh water, and what little there
is has to be collected in pits dug for the purpose; there is
an utter absence of bays and inlets, with the exception
of a few bights not sheltered from the sea wind . . .*

Yet, on the east coast in 1770, the abundance of the land was so striking that Captain Cook could write in his journal: 'The great quantity of new plants etc. Mr. Banks & Dr. Solander collected in this place occasioned me giving it the name "Botany Bay".'

The new land was slowly being revealed, but more than another century was to be needed before it could be said that the 8 million square kilometres of Australia had been explored and much of the inland found to be unsuitable for human occupation. The names of the explorers who braved the harsh horrors of the great deserts and the adjacent country, men like Eyre, McDouall Stuart, and Giles, are a part of a robust saga that has left only a sadness. Where they went few pioneers followed, or are likely to. The hostility of the land that they endured, country so austerely beautiful, from a safe distance, in its huge, hot, red aridity is the same today as it was then, and settlement on any significant scale occurs only when the minerals of the area, the metals that the Dutch thought must be non-existent, are abundant enough to justify the importation of what is, in effect, a complete, air-conditioned, more temperate environment.

White occupation and subsequent settlement took place first along the coastal areas of the country, and then primarily in the east with its higher rainfall, gentler climate, and easier conditions. Here, to the sclerophyll forests with their backing areas of wooded parkland and savannah grassland, came the convicts and first pioneers who also fulfilled an exploring role as they pushed their agriculture and settlement further and further into the rich and seemingly empty land. From what is now the Sydney area the explorer-pioneers fanned out north, south and west, sometimes compelled by the novel conditions they encountered, as when Blaxland, Lawson and Wentworth crossed the Blue Mountains fleeing from caterpillar plagues and drought.

THE FIRST PIONEERS. Once the initial settlement was established at Botany Bay the occupation of the land followed with remarkable rapidity. Free settlers and ex-convicts followed the typical pattern of exploitation and development of a new country, still in innocent ignorance of the destruction they were causing and the damage they would leave to subsequent generations. To their Adam eyes and Eden vision the land was apparently unlimited and there was scarcely a serpent in sight. The result was that the early graziers overloaded the seemingly abundant land with sheep, and the native grasses hitherto only lightly grazed by the native marsupials quickly declined. The process of erosion began, followed by tunnelling, and finally the bare, crumbling, ugly characteristic gullies came out like sores on the land. That was how it was in central Victoria. Elsewhere destruction of a slightly different sort was in the making. Once natural cover was reduced the essentially poor soils compacted and became unable to absorb rain water, which instead ran ruinously, erodingly, over its surface, producing the extensive gully erosion that is widespread throughout eastern Australia.

THE FARMER-EXPLORERS swarm across the empty map of the east, busy as bush flies round a carcass, a legion of indomitable, indefatigable, curious, optimistic men. They were aggressively intent on finding their fortunes somewhere in the huge, rich quietness of the deserted landscape that they were already beginning to call 'the bush'. Soon their names are too numerous to record comfortably and only a few of the more outstanding are left to fix themselves, alongside their exploits, in the memory: Cunningham discovering the Darling Downs and reaching the new convict settlement at Moreton Bay; Hume and Hovell finding the River Murray and pressing on to the huge, natural harbour of Port Phillip Bay; the opulently equipped, and strongly backed Mitchell setting out with the enormous, for those times, expedition of 30 men, 11 drays and carts, 2 boats, 112 bullocks, 17 horses, and 250 sheep, to discover the riches of Western Victoria; only to find the Henty brothers already in occupation at Portland where they had come by boat two years before, from Tasmania.

The story of the settlement of the eastern quarter of Australia in the nineteenth century is the story of the introduction of European farming which was completely alien and unsuitable to the land and climate, and a retreat into the pastoral practices of grazing sheep and cattle. In many parts of the country (e.g. cattle grazing in the alpine areas) these methods have done as much, if not more, damage to the natural environment as the original European methods. Both forms of agriculture have, in their very different ways, gone far towards destroying a unique plant and animal life, and have changed massively the character of the environment, its soil, its cover, its water resources. The east today is not the land that a thinly scattered population of Aborigines knew until white settlement began in 1788.

Not all the early pioneers were ignorant of the possible consequences of their activities. 'Count' P. E. Strzelecki, contender with his countryman, fellow-Pole Lhotsky, for the honour of the first ascent of Australia's highest mountain, reported on his exploration to Governor Gipps, and said:

> *The drought however, here in New South Wales*
> *seems to me to have an additional cause to that or*
> *those which elsewhere occasion extraordinary dryness of*
> *the soil: namely the alteration which colonisation*
> *impresses on its surface; the herbaceous, high and thick*
> *plants; the continued forest; the underwood; the brush,*
> *which so well clothed the crust and sheltered the*
> *moisture, have disappeared under the innumerable*
> *flocks and axes which the settlers have introduced. . .*
> *.The increase of grazing flocks, which in many cases*
> *overstock the pasture, inferior agriculture, seeds more*
> *inferior still, are alone progressing from day to day,*
> *enervating the surface, and assimilating its best parts to*
> *the rest already sterile.*

Strzelecki's condemnation of the crudities of colonial agriculture of that age and his suggestions for an intelligent form of farming have a modern conservationist ring to them. Then, all the time, as now much of the time, no one listened, or if they listened, brushed aside the arguments in favour of short term advantage or profitability.

The first farmers came to a wide variety of lands. In the great flatness of the Mallee region of Victoria and South Australia they rolled and burned the scrub and put the land to cereal growing without understanding the nature of the light textured soil, and its capacity for accumulating the little moisture provided by the sparse annual rainfall. The result was serious wind erosion and the rapid accumulation of salt in the soil in poisonous proportions in many areas. In recent years new farming methods have been worked out by agricultural scientists, much damage repaired, and the land stabilised in a form very different from that of the tens of thousands of years of Aboriginal occupation.

In the rolling hill country of Victoria, with its higher rainfall, tall-timbered valleys, and the seemingly endless dry bush of the slopes, farming took a severe toll of the delicately balanced natural environment. The soil was poor, overlying thick yellow clay; and heavy grazing, the clearing of trees and their killing by ring-barking led eventually to tunnel and gully water erosion throughout much of the area. A similar story of water erosion can be told of the Darling Downs in Queensland and the western slopes of the hills of New South Wales.

In the alpine regions of Victoria and New South Wales grazing, burning, and clearing on a large scale quickly altered the delicate balance dictated by the harsh environment and narrow temperature range. For most of the nineteenth century the alpine areas were free grazing country for cattle, sheep and brumbies. It was to be well into the twentieth century, and thanks to the need to preserve the catchment areas of the several alpine hydro-electricity schemes, before substantial control was established, the damage halted and partial rehabilitation begun.

In these areas where water is relatively plentiful, but the harshness of the conditions is severe, quite small interferences with the environment can produce massively damaging consequences. The summer grazing of sheep and cattle and the burning of pasture to produce new growth brought about widespread changes so that the region would now be almost unrecognisable to the Aborigines who for so many hundreds of centuries climbed every summer to the mountaintops to feast on the unnatural summer snow of the huge swarms of plump White Bogong moths that gathered there.

The alpine region is terribly small in proportion to the rest of the continent and therefore especially precious. There are only 6500 square kilometres of land over 1400 metres in height, which amounts to less than 0.5 per cent of the total area of Australia. The central area is now the Kosciusko National Park, and the high heartland of 180 square kilometres has been proclaimed as a wilderness area within the park. Grazing on the uplands has been prohibited for some years but is continuing in the foothills. The

Australia's Alpine Regions

latest threat to the area lies not in agriculture but in its exploitation by commercial elements, catering for tourists in both winter and summer, with the high possibility of the domestication and Disneyfying of the region.

The land remains in all its wonderful variety of shape and cover and climate but altered almost beyond recognition in the almost 200 years of white occupation and management. The only true wilderness remaining in the whole of eastern Australia is in south-west Tasmania, and this, with the flooding of Lake Pedder and the carrying out of hydro-electricity schemes in its northern part, is now considerably modified and further threatened.

The last vestiges of the old thoughtless attitudes of the first pioneers are still alive and painfully evident across much of the countryside despite the activities of conservationists, bush walkers, campers, and all the concerned people in Australia who look to the bush and the mountains for refreshment and renewal. These people, young and old, are the first wave of new pioneers who will go to the land in the future, as many of them do now, intent on preserving and conserving instead of exploiting and destroying. The new pioneers cannot undo the damage of the past, but they can help to reclaim the reclaimable, and preserve all of what remains.

Further Reading

D. J. Mulvaney, *The Prehistory of Australia,* Penguin Books, 1975.
J. Flood, *Archaeology of the Dreamtime,* Collins, 1983.
Wade and Hughes, *Exploring Ancient Australia,* Horwitz Grahame, 1988.
C. H. and R. M. Berndt, *The Aboriginal Australians,* Pitman, 1983.
B. Attwood, *The Making of the Aborigines,* Allen and Unwin, 1989.
Feeken, Feeken and Spate, *The Discovery and Exploration of Australia,* Nelson, 1970.

NATURAL EQUIPMENT

*The swag is usually composed of a tent 'fly'
or strip of calico (a cover for the swag and a
shelter in bad weather . . .) a couple of
blankets, blue by custom and preference . . .
and the core is composed of spare clothing
and small personal effects.*

HENRY LAWSON

Houses are static defences against nature and the elements. Tents and fabric shelters are really a form of extra clothing, but a form inside which you can move about. They are not houses nor substitutes for houses. This is an important distinction which is often, and too easily, forgotten. Houses are a rejection of nature; tents and fabric shelters are the apparel you wear at various times outdoors to enable you to live continuously and acceptingly with nature. The Aborigines built no permanent dwellings, understanding as they did that in Australia integration with nature is inseparable from movement. Their shelters were, and in many cases still are, temporary and mobile.

In the same way the equipment you carry is *not* furniture. Like the shelter clothing which gives you mobility, your equipment is to enable you to live and move about in the natural environment with confidence and in reasonable comfort. Here, as in art, the rule has to be: the less the more. An abundance of objects of the sort you are accustomed to doesn't necessarily make for comfort and efficiency, and usually results in repetitive complications. Weight is also an important factor, regardless of the sort of camping you intend to do.

In general it is safe to say that the more objects you lug along on your camping expeditions, the less you will encounter the nature for which you set out. An excess of objects gets in the way, interposes a screen of worry and preoccupation between human beings and their natural surroundings, obscures the different view they are seeking. The question is: what can be done without? Indispensability is the first quality required of any piece of camping equipment.

TENTS AND SHELTERS are just as subject to this criterion as any other piece of equipment, and more so than most because of their bulk and weight. If you are going walking in hilly, open country in fine, settled, summer weather and are indifferent to insects or happy with the protection given by chemical repellents, then you can probably manage without a tent at all. This is an ultimate, in that you can move further more freely and happily than in any other way and a prime example of 'the less the more'.

If you want to take out insurance against the odd shower or heavy mist you can invest very cheaply in a sheet of plastic weighing only a few grams. (Be sensible, get one of reasonable thickness, otherwise it will only rip itself into useless shreds as soon as you try to use it.) Use this folded under and over you as a plain cover and groundsheet; slung over a branch and tied, or weighted down with stones to make a simple A-shape shelter; hung, on a length of nylon cord from your pocket, between two trees; or in any way which strikes your fancy as affording the sort of protection you need. In the United States they now have what is called a tube tent. This is a cylinder of plastic that can be strung between two trees and gives all-round protection, except at the ends. Tube tents are miserable places to be in wet-and-windy conditions, but handy otherwise.

Leaving plastic sheets and the ingenuities of improvisation, one comes to an area of controversy. Informed, tradition-following opinion says that, in general, tents for Australian conditions should be made from japara, which is a very high quality cloth of finely woven cotton. Japara can be treated to make it waterproof, but still allows air and moisture to pass outwards, so that the moisture

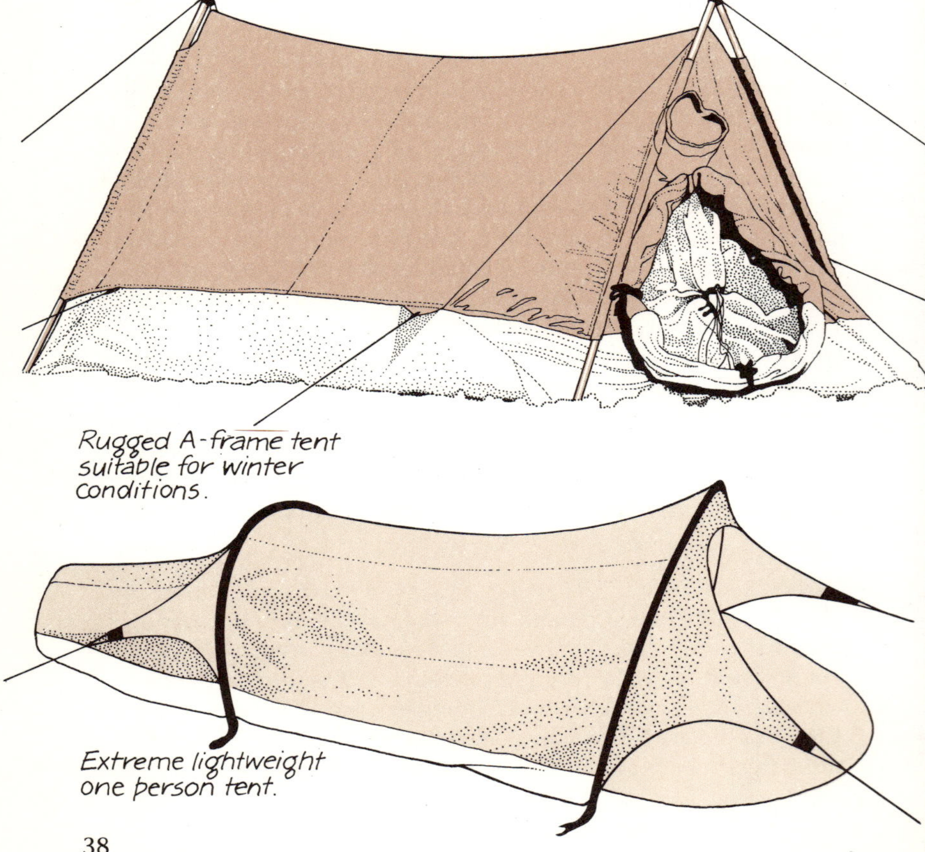

Rugged A-frame tent suitable for winter conditions.

Extreme lightweight one person tent.

breathed out by the tent's occupants, particularly when asleep at night with the tent closed, can get away.

Another, and growing, school of thought, is that *proofed nylon is unbeatable as tent material,* being cheap, making a considerably lighter tent than japara, and being almost completely proof against water. The drawback of nylon is that it does not 'exhale' the moisture breathed out by the tent's occupants, so that the insides of nylon tent walls are soon running clammily and uncomfortably, with condensation. In recent years these nylon tents have been designed with 'ventilators' to allow condensation to escape, but this has proved only a partial solution to the problem. Nevertheless the advocates of nylon as a tent material claim now to be satisfied and point out that japara, even when chemically proofed, is not entirely waterproof. This means that heavy rain, or rain accompanied by strong wind, soaks through the fabric, misting the tent interior and dripping on the occupants wherever the tent walls have been touched.

This is perfectly true, and the nylon supporters have a point here, but it is one that is swiftly answered by the japara supporters who say that for rough weather conditions a tent fly can be used. A tent fly is an outer cover of japara or nylon held several inches away from the actual tent wall. It functions like an umbrella against the rain while allowing air to circulate and the inner tent to 'breathe'. The fly adds to the weight to be carried but is almost indispensable and well worth the extra effort in areas where wet weather is persistent.

Possibly the best material of all for tents is Gore-Tex fabric. This is a thin plastic sheet, with millions of perforations to the square centimetre, which is laminated between two strong light fabrics. Gore-Tex is waterproof and also has the capacity to 'breathe' out condensation forming in the tent. A single layer Gore-Tex tent requires no flysheet and is proof against high winds. It is best not to cook inside a Gore-Tex tent, but if cooking inside is unavoidable the tent doors must be left open, otherwise there is the risk of deadly carbon monoxide poisoning.

In the generally mild and dry conditions of Australia a lightweight japara tent and flysheet is probably the best compromise. In alpine regions in winter, with the possibility of snow and high winds, a high quality double thickness nylon or single thickness Gore-Tex material should suffice. For snow conditions there are a number of specially designed and constructed tents on the market, usually dome or tunnel shaped.

Larger tents are made of different and heavier materials, are ventilated differently, and don't present the same difficulties. Small tents are simple in design and come in limited shapes. Two large clumsy people who like to stretch should go for an oversize tent.

A GROUNDSHEET OR FLOOR for the tent is essential. This can be made of nylon or plastic, sewn in (mandatory in snow), detachable by zip or tabs, or completely separate. The detachable groundsheet that is designed for, and snugly accommodates itself to, the tent is the best of these options.

Larger tents come in a wide variety of shapes, sizes and materials. Probably the best on the market today are those based on European designs and hung on lightweight metal frames. These frame tents commonly have one or more zip-up compartments suspended from the frame as separate sleeping areas, and large netting windows which can be covered by roll-down translucent panels against wind or rain, and an outer roll-down panel of tent material at night. Despite their size and substantial character these tents are *relatively* light and occupy only a small space when transported. The fabric 'breathes' and the roofs and skirts of the tents are made of quite heavy plastic, as are the bases of the sleeping compartments.

With larger tents mobility is lost as motor transport is needed to move them, and they are hardly worth the trouble involved in pitching each night, even if they were properly portable. Once the small tents are left behind, a different concept of camping, the fixed camp, a base for leisurely holiday or exploration, is involved.

C ARE OF TENTS. Tents should never be washed. Mud-splashes soon dry off and are then easily removed. The risk of damage or destruction by fire or sparks is a very real one. Don't cook in a tent except when absolutely essential. Always make sure that the tent is well aired before storing it away between trips. Temporary repairs to small holes and tears can be made with adhesive tape or bandage.

P OLES AND PEGS. Most tents come with alloy poles or frames which break down into convenient short sections for carrying. Many large modern tent frames are connected in sections by wire springs which hold the joints together and greatly simplify the erection of the often quite complicated framework. If there is an urgent need to save weight, and the camping is to be in areas where there is plenty of bush, tent poles can be left behind and branches utilised instead. This is true only of small overnight tents and the weight saved is usually so small as to be hardly worth the trouble.

Pegs also accompany a new tent and are usually made of five millimetre diameter aluminium wire which is adequate for most conditions where a small tent is pitched. Larger tents often come with similar lightweight pegs and it is worthwhile, since transport is involved, to carry an extra heavier set for use in awkward ground or hard conditions. In sand or snow the only really efficient pegs are those made from angled aluminium or galvanised steel strip.

Campers, like sailors, should always carry a spare length of suitable strength cord with them. The cord takes up a negligible amount of space and is invaluable for emergency guy rope repairs as well as having innumerable other uses.

TENT AND FLY MAKING. Making your own tent at home can save money but depends on whether a suitable sewing machine is available. Both japara and nylon are reasonably priced. Home designs are not usually satisfactory. Tents are such an old form of protection that good innovations are now usually technical and concerned with materials (e.g. nylon) and small gadgets (e.g. zip fastenings). If a tent is made at home a good pattern to copy is essential, and preferably a pattern of a tent you know, have used, and learned to trust.

The fly for a tent is quite easily and cheaply made at home with the added advantage that it can incorporate whatever provision you want to make for the idiosyncracies that your tent has shown in use.

SLEEPING BAGS. The Americans macabrely call sleeping bags *mummy bags*. This is a grotesque and quite superficial description. A good quality sleeping bag is a soft, airy and delightful article, warm and reassuring and quite beautiful. Properly crafted sleeping bags have an active life in them that is immediately recognisable. Inferior bags feel, and are, inert, dead, and therefore inefficient and wasteful.

The life in a good bag comes from the down that fills it and that traps the air to provide the insulation that ensures warmth. For camping, a down or superdown filled bag is an essential unless your intention is to camp only at the height of heat waves. For caravanning or trailering or sailing, a bag filled with a synthetic mixture, or even wool or cotton, can be used. Like tents, sleeping bags have to breathe, so they can't be waterproofed without your own sweat condensing clammily on to you. The higher the quality of bag and down the more you have to pay. Wool or cotton filled bags are cheapest, synthetic filled bags are rather more expensive, and down and superdown bags are the most expensive. Bulk is important. Down bags will compress into quite small carry bags; the others take up considerably more space and are more difficult to pack.

Much of the efficiency of the sleeping bag depends on the way the filling is distributed. Ordinary quilting is not as good as walled (or box) quilting, and leads to considerable heat loss through the zip area and seams. Walled quilting prevents this happening. Ordinary or sewn-through quilting makes pleasant patterns on the outside of the bag which are cold lines because they offer no insulation all along their lengths. Walled quilting, using box patterns of broad tape as the walls, completely obviates these cold areas. There are other structure patterns of quilting, such as the tube, slant tube, overlapping tube, and laminated.

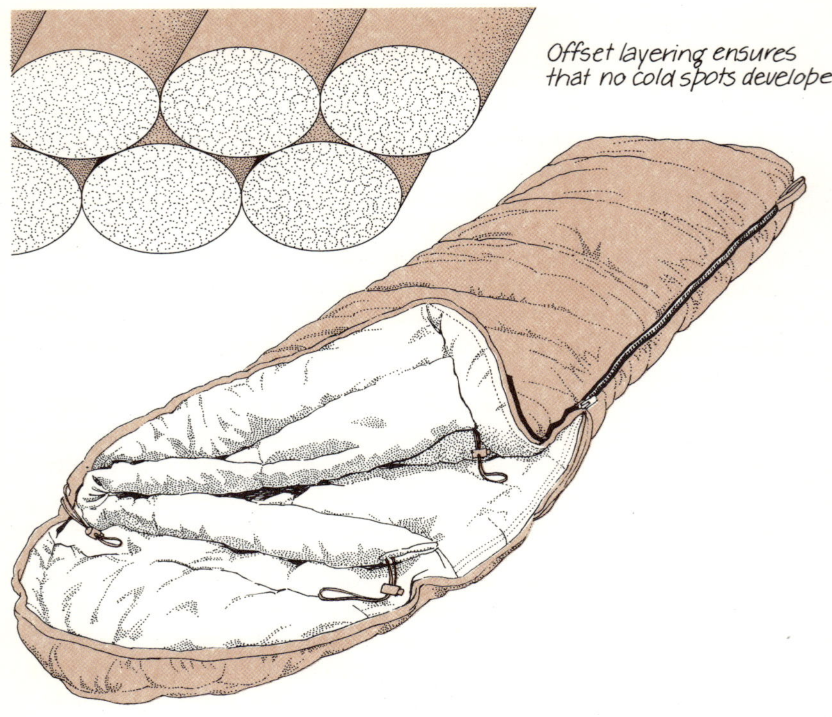

Today, sleeping bag shells are usually made of untreated nylon or fine cotton. Nylon is light, can 'breathe', is strong, and has a slippery surface which some people find offensive but deal with by carrying a light cotton inner bag for lining. Nylon is also highly susceptible to damage from sparks.

Shape and size of sleeping bags are important to their insulating qualities. There are three basic shapes: rectangular, barrel, and mummy. Barrel and mummy bags are tapered towards the foot, and have tailored or circular hoods closed by a drawstring. The mummy claustrophobic for some people. Little or no warmth is allowed to escape but neither, often, does the occupant. The circular, drawstring-operated hood of the barrel bag is almost as efficient and much less claustrophobic, and the body of the bag is not so close fitting so that some movement is possible. Uneasy sleepers, active in their dreaming, are best catered for by the rectangular, or only slightly tapered bag.

The size of the sleeping bag is important. Bags are made in a variety of lengths which are not quite what they seem on closer examination. The body is longer when lying down, the bag is shortened by passing over the bulk of the body, and room has to be allowed for turning, flexing, and stretching. It follows inevitably that a bag marked over 2 metres in the shop will not provide a comfortable place for a person of over 2 metres in height. On average, for barrel and rectangular bags, the extra length should be 20 to 25 centimetres and for mummy bags 7.5 to 10 centimetres up to 1.5 metres in height, and 25 to 30 centimetres up to and over 2 metres in height.

Sleeping bags made in Australia and of quality adequate for bushwalking can be found in three main categories. These are bags intended for high altitude/snow conditions, generally cold conditions, and the mild conditions encountered by the occasional fine weather walker. The bags in these three categories are filled with superdown, down, or feather down. Bags containing other filling materials are not suitable for bushwalking.

Without transport, weight is an important factor. A good bag is worth its weight in comfort and succour, especially at the end of a bleak, cold day, but in weights of 2 kilograms and over is not really practical for walkers. Fortunately, first class sleeping bags can be obtained that weigh from 1.6 kilograms upwards, and there are a number of top class sleeping bag manufacturers in Australia.

Good sleeping bags are expensive but will outlast cheaper articles many times over and *always*, if properly treated, provide immeasurable warmth and comfort. Couples who want to maximise their sleeping bag investment would do well to pay a little extra and get matching bags with matching all-round zips. These bags can be used separately, zipped together to make a double bag, and combined, when not in use for camping, as a luxury double bed quilt at home.

Zips on sleeping bags can be a blessing or a curse. Zip failure can lead to discomfort on chilly nights, and laborious and usually unsatisfactory repairs. Regardless of the bag that is bought, make sure that if it has a zip, it has a good, robust, smooth working one.

CARE OF SLEEPING BAGS. A lightweight detachable cotton lining saves wear and tear on your bag. When not in use hang or otherwise arrange your bag so that the greatest possible expansion of filling is allowed. Always handle your bag carefully and make sure that when it is laid down there is a protective material (groundsheet, airbed . . .) between it and the ground. Bags are best dry cleaned, but they retain the poisonous cleaning vapours and must therefore be very carefully aired afterwards. Dry cleaning does not extend the life of your bag; quite the reverse.

AIR AND OTHER BEDS. Sleeping bags insulate the body from the outer air whenever the down can expand or 'lift' to its fullest extent. Obviously the down crushed flat beneath the sleeper is ineffective either as insulation or cushioning. Even when the sleeping bag is placed on the groundsheet of a tent the heat loss is substantial and the discomfort, after only a few minutes, unpleasant. In the bush it is usually possible to improvise a quite satisfactory bed of brushwood and other ground materials which will provide some degree of insulation and comfort, but this is not necessary as well as being yet another small incision into an already fast diminishing bush. A better alternative is available.

An increasing number of hardy, strong-boned young people are solving the bedding problem with sealed cell foam. This comes in body lengths and widths and is 1.25 to 1.95 centimetres thick, and weighs about 500 grams. The pad rolls up into a very small bundle, and is watertight. These foam pads are primarily for insulation and are very effective. They are not particularly comfortable even in their thickest sizes. Inflatable foam mattresses are an improvement and widely available.

A three-quarter length air mattress offers more cushioned comfort, weighs only about a kilogram, and rolls up into a small parcel for carrying. As insulation against heat leakage to the ground, air beds are inefficient. Convection currents within the bed cells exchange the cold air of the ground for the warm air from the sleeper on the bed's surface. Air beds require daily inflation or deflation, which takes time, and are subject to slow leaks and punctures if not handled with proper care. Rubberised fabric is quite easily repaired (carry a kit and spare plugs), but with commonsense usage punctures should be rare.

Air beds are also made in plastic which has the virtue of extreme lightness but the vice of extreme fragility. Plastic punctures and tears

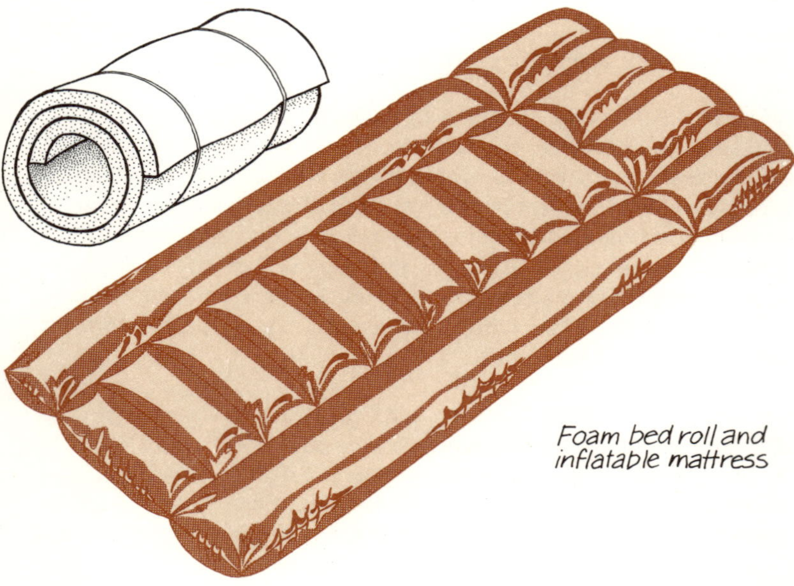

Foam bed roll and
inflatable mattress

very easily and can be difficult to repair. All in all, plastic is an unsuitable material and will usually cause far more trouble than it is worth.

KNAPSACKS, RUCKSACKS, PACKS. The convenient carrying of even a few necessaries such as lunch and spare sweater involves the use of a backpack of one sort or another. On short walks of a day or less a knapsack, hung on the back from the shoulders, is usually adequate. Knapsacks have no framework and are simply bags, often of strong nylon, or other synthetic material, with shoulder straps. Their only inconvenience to the wearer is that in warm weather or when making considerable physical exertion they cause unpleasant sweating where they rest against the back. Even the simplest knapsack performs the necessary function of keeping the hands free, carrying the small load where safest, and keeping the load from snagging or catching against rock or vegetation.

A rucksack is a large knapsack with pockets on the outside and with a steel or alloy A-shaped frame attached. It performs the necessary pack function while holding the pack itself away from contact with the back so that excessive sweating does not occur. In the past A-frame rucksacks have been produced with the greatest volume of the main pocket at the bottom, disregarding the fact that loads are better carried high up and close in to the shoulders rather than at the back at waist level. Modern rucksacks now have bags designed to be narrow at the bottom and with the greatest volume at the top, which has improved their efficiency and comfort greatly. This type of pack is designed to carry considerably heavier loads than the knap-sack and can be used for heavy or bulky day walk loads or for short camping trips. Its chief disadvantages are that the outside pockets easily snag or bump against any obstructions, and that it is a compromise in terms of design efficiency. Science notwithstanding there are still a number of people who would not carry any other sort of pack.

The modern design pack that has swept almost all before it is the metal H-frame, which is really an inverted U shape with curved crosspieces and webbing to keep the bulk of the frame from direct contact with the back.

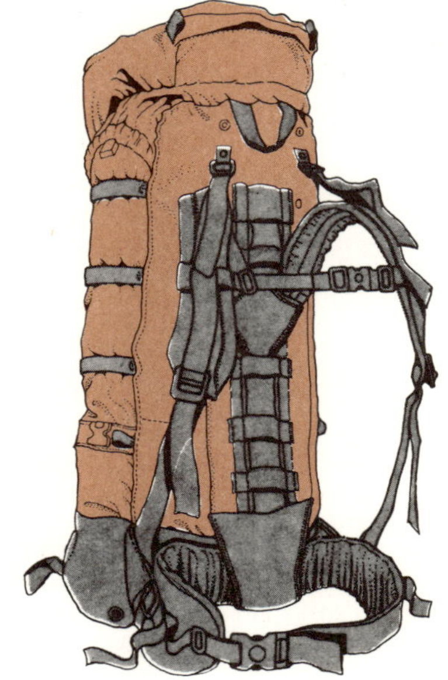

Modern internal frame pack well supplied with straps for comfortable load adjustment.

The H-frames come in several sizes and usually have both shoulder straps and waist straps to keep the load close in to the back and carried by powerful muscles of the hip area. The bags for H-frames can carry up to 30 kilograms and are made of a variety of materials, though heavy nylon which is durable and light and can be made water repellent is now being favoured. Water-proofed bags cannot breathe and anything inside them that can sweat will sweat on to the other gear. Bags are best reinforced on the bottom. Some walkers carry waterproof covers for their bags. Extra pockets are best on the back of the bag, in preference to the sides where they may stick out and catch on trees or rocks.

Bags can be bought which have a variety of rings and loops and straps from which almost everything but the kitchen sink can be hung. If you have no particular use for these gadgets, leave them off. This will make walking pleasanter and easier and avoid the distressing and comical Christmas tree or candelabra effect, though with some of the larger, taller packs which come as high as the top of the carrier's head, this sort of effect is unavoidable.

Modern packs can be bought in a variety of sizes and it is best to buy a slightly bigger bag than is likely to be needed. There is a temptation to go for a size very much larger than is needed, but this should be resisted. A lot of spare space is a danger, because it invites filling with unnecessary gear that weighs unduly and never gets used. A towering, tightly packed bag may look impressive but it still has to be carried, and it is a well-known fact that weight increases in direct ratio to the distance covered. A good general rule in estimating loads to be carried is to make sure that the weight does not exceed one-third of the body weight of the carrier. Using this formula about 25 kilograms is the *maximum* sensible weight for a 76 kilogram man, and about 19 kilograms the maximum for a 58 kilogram woman. Expert campers, like philosophers, agree with Plato who said in effect that a man can only be considered rich in proportion to what he can manage to do without.

Further reading

Paddy Pallin's Bushwalking and Camping, Sydney, 1985.
Chapman and Chapman, *Bushwalking in Australia,* Lonely Planet Books, 1988.
R. Graves, *Australian Bushcraft,* Dymocks, 1984.
Jeff Carter, *The Australian Explorer's Handbook,* Hodder and Stoughton, 1987. (Mostly motor touring, includes outback and desert regions)
The indispensable guide in Australia is: *Equipment For Bushwalking and Mountaineering,* Melbourne University Mountaineering Club, 1982.

CHAPTER 5

THE UNWELCOME MIGRANTS

This is adding insult to injuries.
EDWARD MOORE

When the first white people came to their involuntary exile in Australia they found a country only meagrely provided with natural foods. The 300 000 Aborigines, spread across the huge area of the continent in their ceaseless movement, were probably the largest number of people the country could support in its natural state. The first convicts and their escorts learned the bitter lesson of hunger in the first years of their settlement's existence. Captain Phillip, the first governor, was of the opinion that 'perhaps no country in the world affords less assistance to first settlers'. This was not surprising in a land in which the native grass killed the few sheep (precious few had survived the long voyage) who ate it, and the newly imported cattle escaped and wandered off into the bush beyond recapture. An attempt to grow European vegetables and grain failed miserably in the sandy soil. Two years passed before the next supply ship arrived from Britain and the settlers had to live out the time on protein-deficient short commons.

The new country seems to have been as intimidating psychologically as it was hostile materially. Kangaroos were frequently shot but their flesh was disliked, and fish were reported to be scarce, which in the light of subsequent experience on the coast of New South Wales seems unlikely, and possibly, a reflection of the state of psychological beleaguerment the settlers went into as time passed and their skimpy rations steadily decreased.

The first settlers learned quickly. Almost everything of use to them had to be imported and introduced by trial and error to the new soil. Today, almost 200 years later, it is possible to go from north to south and east to west across Australia and be sure that almost every useful domesticated crop or animal seen on the way is an import. Without adapted European farming methods and animals and plants it would be impossible to support the existing population.

Once farming became established the flow of imports from Europe grew into an indiscriminate flood carrying as much with it that was harmful as was good. The innocence, naivete and homesickness of the early settlers combined to bring in the harmful plague of blackberries to Tasmania as early as the 1840s. They were imported because, at the time, an attempt was being made to acclimatise the curassow, a big turkey-like bird from South America, which was said to thrive on a blackberry diet! The blackberry soon reached the mainland by one means or another, and is not completely eradicated today.

ACCLIMITISATION SOCIETIES throve like the wretched blackberries during the nineteenth century. The long-undisturbed continent was proving a generous host to so many useful crops and animals that bizarre and outlandish imports from all over the world were attempted in that first century's blazing enthusiasm. Secretary birds, famous for their snake killing prowess, were brought from South Africa, but failed and died out. Mynah birds were imported from India in 1862–63, together with English sparrows, to keep down caterpillars. Both birds and caterpillars prospered. Queensland got sparrows, thrushes, blackbirds, rooks, starlings and larks in 1869. Later an attempt was made to introduce pheasants, but this failed. Pheasants are now bred for shooting in Tasmania. An advertisement in the *Australian* in the spring of 1975 headed Pheasant Shooting 1976, offered: Opportunity for selected Game Hunters to acquire a 'Gun' for the 1976 season . . . (Limited 'guns' available). Around the mid-nineteenth century, ostriches arrived and some still survive in South Australia.

Odd as anything ever attempted was the introduction of alpacas and llamas, domesticated animals from South America. The llama is one of the most useful of all-purpose beasts. It is eaten and relished, used as a work animal, and produces a good quality wool. The llama is notorious amongst animals for its ability to spit out, accurately, a foul smelling expectoration, the traces of which are difficult to cleanse. Another of its little idiosyncracies is its ability to gauge the weight loaded on its back. Any load much over about 15 kilograms produces a stubborn refusal to move. Its dung is an important natural fuel in Peru and Bolivia. The alpaca, a close relative, which was revered by the Indians, was used only for the fine wool it produced. Both llama and alpaca are unprepossessing in appearance, looking like varieties of large, long-legged sheep, with distinctly superior expressions on their fine-drawn faces.

The first alpacas were brought to Australia by a Mr C. Ledger. In the mid 1850s the South American natives were hostile to the idea of exporting alpacas. Ledger bought animals and gathered them together secretly to drive them into Chile. This involved crossing the Andes at about 6000 metres on dangerous tracks through storm and snow! Before this he had brought his beasts through a countryside of hostile natives, past whom he sometimes had physically to fight his way.

The llamas and alpacas were accustomed to the high, arid, intensely cold conditions of the Andes. In Australia they encountered heat, a too rich diet, disease and damp, and could not survive for long. It was intended that they should be taken to the high country of the Snowy Mountains, but none ever reached there.

Camels were brought in during the early 1840s, and the first attempts to introduce trout and salmon were made at this time. Trout took on and prospered eventually because of the absence of enemies, the excellent cover in the rivers and streams, and the richness of the feed available. The Atlantic salmon never succeeded in establishing itself, but the European carp, one of the most damaging and useless of freshwater fish, succeeded in a big way. Large scale attempts to exterminate this fish have failed and it is now widespread and polluting new water every year.

The small as well as the large figured in the beaver-busy (there is no record of an attempt to introduce beavers) activities of the ubiquitous Acclimatisation Societies. In New South Wales an attempt was made to breed cochineal insects which are the source of a valuable red dye. The cochineal insect is a parasite of the prickly pear! The cochineal insects died out. Silkworms, too, were introduced at considerable cost, but because of the absence of abundant cheap labour they did not become the basis of an industry. Relegated to the status of pets for children, they nevertheless attracted the attention of an educational reformer of the times who suggested that orphan and reformatory children would be better employed breeding silkworms and producing silk than wasting their schooldays on more orthodox forms of education.

The Acclimatisation Societies either died out or turned themselves into zoo providers by the end of last century. There were few things that the societies did not attempt. A short, selective glance at some of their efforts shows that we owe to them the coarse and unpalatable carp, the tench, the roach, the redfin, and the European carp, as well as the gambusia fish (its Asian name means 'worthless', and it is), the raven, the crow, the cattle tick (brought in with zebu cattle), and many other imports of fantastical selection and doubtful utility. The enthusiasm of the Acclimatisation Societies was indiscriminate, their foresight negligible, and the consequences of their uninformed activities, in general, highly damaging to the ecology of the country and the comfort of the inhabitants.

INCIDENTAL, INADVERTENT AND ACCIDENTAL IMPORTS. A number of varieties of prickly pear was brought into the country for several reasons. Prickly pear accompanied the cochineal insects which failed to survive, but it was also imported for stock feed and to make hedges. The prickly pear escaped and ran through the country of southern Queensland and New South Wales like wildfire. Like the blackberry in the south it throve in its new environment, growing up to 2 metres in height and completely dominating the land it infested. Its eradication seemed impossible for many years though a variety of methods was tried. Poison and poison gas were used, but nothing proved adequate. By 1925 12 million hectares of country were covered by its almost impenetrable growth.

Lantana sp.

It was at this time that the first eggs of a moth, whose larvae attack prickly pear, were imported from the Argentine. Breeding was quickly expanded in laboratories and eventually 3000 million eggs were sent out free to landholders. In Queensland the menace was eradicated, but the moth, which needs a dry, warm climate, could not survive the climate of New South Wales where the battle to eradicate prickly pear is still continuing.

Rats, mice and flies came to Australia as the inevitable incidentals of maritime communication. Ships' rats made their way ashore, mice nested amongst the springs of imported chairs and in crates and boxes, and the packing of straw and shavings and sawdust. Illegal immigrants, they entered secretively a country already populated with more than fifty varieties of mice and rats. Occasionally the migrants swarmed in plague proportions and tried to take over. During a plague of house-mice in Victoria in the First World War, poison was laid one night around some bagged wheat. The next morning it was found that 7 tonnes of mice had been poisoned!

The importation of the European green blowfly was accidental, but it badly damaged the sheep industry and made life near to impossible for sheepmen, who had to try to cleanse the sheep of the parasitic maggots that resulted. The disease-spreading home-fly came in incidentally with the ships and a bit of colonial mythology, which made the home-fly the adversary and conqueror of the native blowfly, had settlers trapping home-flies and carrying them to their properties.

Pigs were brought in willy-nilly almost from the beginning of settlement, and were soon running wild in all directions. They are still probably the most dangerous quad-rupeds in the country. A cornered boar is almost as dangerous as any animal alive. Wild pig is still plentiful in eastern Australia, and is well known for its vacuum-cleaner-like efficiency in

Cane Toad (Bufo marin

consuming every last portion of its victims.

Imported domestic dogs were quick to get out of hand and start hunting in packs. The streets of the early settlements were often unsafe to walk at night because of the many dog packs roaming the shadows. Dogs bred for hunting and made superfluous by falling prices were turned loose without regard for the consequences.

Donkeys, like camels, were turned loose to forage for themselves when motor transport was introduced. They bred up quickly in the north of Australia but both animals have proved unexpectedly useful in recent years. In a coals-to-Newcastle contract, Australian camels have been sent to the Middle East to improve the stock of the local racing breeds. The donkeys, which are shot in tens of thousands, are a part of the growing and thriving pet food industry.

Domesticated farm animals have escaped and run wild in their thousands. Cattle and goats have survived and bred up in large numbers until recent years when the cattle were gradually rounded up for the American hamburger trade, and it was found that there was a large market for goat meat in South-East Asia. Horses, paradoxically, run most intelligently wild and evasive. They are called brumbies in Australia after a soldier-settler of around 1800, Private James Brumby, who farmed and bred horses, but let them go free when he moved off to Tasmania.

There have been times when wild horses have assumed plague proportions. About 1870 settlers resorted to poisoning, fencing off water holes which left the horses to a lingering death, and round-ups which ended in the horses being shot. Disposing of the carcasses was a problem since the fuel required for burning them was unavailable. Eventually a practical, but macabre-minded genius invented the 'three mile range' method, in which the horse was shot, with considerable precision, in the heart, after which it could run for about three miles before dropping dead. Thousands of horses still roam free in the wilder areas but their numbers are kept down because the flesh is valuable for the pet meat industry and for export to Europe for human consumption, though this latter gourmet trade is unofficial and unrecognised.

DINGOES are undoubtedly the oldest imported inhabitants. They are believed to have been brought in by Aborigines some few thousand years ago and were usually kept more as domestic pets than hunters. There was little to stop them running wild in the gentle abundance of the bush where there were marsupials to hunt and no significant enemies. By the time of white settlement they existed in very large numbers all over the country and they had, briefly, a field day amongst the sheep and lambs brought by the settlers. The long centuries of freedom from threat had made the dingoes unsophisticated and unwary in the presence of men and they were hunted and poisoned in a huge slaughter that went on for many years as the settled areas slowly extended inland from the coasts.

Dingoes have bred up at intervals intermittently, ever since. Scalp bonuses for dingoes are still paid in some areas, but since the 1960s poisoned baits have been dropped, indiscriminately, in wild areas where dingoes are known to abound, with unguessable results amongst the native creatures and birds who are meat eaters, and the possible destruction of rare and irreplaceable wild species. There are many dog-fences patrolled by doggers all over the country, but these are easily breached by floods and sand and large animals like kangaroos. Dingoes thrive all along the dividing range and would quickly move down to richer land if not kept under strict control. They will attack calves and sheep, but in the wild take kangaroos, emus and rabbits when they can, and lizards, birds, eggs, roots, frogs and water birds when they can't.

Dingoes are handsome and intelligent creatures. They cannot bark, but they can howl, and doggers have learnt to imitate the howl so perfectly that they can call dingoes within range of their guns. This doesn't always work because the dingo may call back and expect the dogger-howler to come to him. The Simpson Desert is the focal point of a mysterious migration of dingoes that occurs irregularly over the years. The dogs move in from as far as 300 kilometres to the east, and reappear singly on the return in emaciated and weak condition. Their source of interest in the Simpson Desert is not known.

A HUNTING WE WILL GO was the English music still ringing in the ears of some of Australia's earliest settlers. The hunting of imported deer was taking place by 1820, and the quarry of the Sydney Hunt a few years later was the dingo. In Tasmania about this time the fast-running forester kangaroo was the quarry which was hunted and it was pursued with a full pack of hounds. Foxes seem to have been imported first around the middle of last century. By the 1880s they were well established in Victoria and South Australia. It was nearly the end of the century before foxes were reported in New South Wales, and within the first decade of the twentieth century bonuses were being paid for scalps. Southern and central Queensland had established populations by the 1920s and Western Australia by 1917. The first foxes were caught in the Kimberleys of the far north-west in 1943. Disease became widespread in the early 1950s and the fox population declined sharply, only to rise steeply again as the survivors recovered and recommenced breeding.

Research has shown that foxes live chiefly on carrion sheep and lamb, and rabbit. Only an occasional 'rogue' will pull down young lambs. As well, foxes are scavengers consuming dead animal remains of all sorts, whether above or below ground, and supplementing their diet with small live creatures like birds, frogs, mice, grubs, crickets, grasshoppers, fish, butterflies, and young emus. They will also steal birds' eggs. To obtain fish they wait for a heron to catch a fish in its beak then harry the bird until the fish is dropped. They will snatch fish from shallow pools themselves. An unexpected and sinister aspect of the presence of foxes in the country was the discovery that they play an important role in the life cycle of the tapeworm which has caused export meat to be rejected.

Despite the fact that foxes are quite at home in very dry country, and are numerous on the Nullarbor Plain, they have not penetrated into the far north. The two most common forms of control are shooting and poisoning. Foxes will dig up buried baits which remain out of reach of other animals and birds, and 'spotlighting' at night from motor vehicles has been found to be the most effective way of shooting.

HARES were first introduced successfully in Westernport, Victoria, in 1859. The busybodies of the Acclimatisation Society of Victoria released four hares in 1863, and continued to import and release them for another ten years. Up to this time hares were protected under the Victorian Game Act which specified a fine of £5 ($10) for taking hares out of season, that is from August to December.

Like so many other imports, the hares prospered and multiplied on the mild climate and abundance of food. Around and for some time after the turn of last century they were in such nuisance numbers that payment was offered for scalps and was made for over 100 000 in 1908. Since this period they have remained within quite easy control of farmers and pastoralists.

The coursing of hares by greyhounds, which had begun illegally in Victoria in 1870, was legalised in 1873 and encouraged the importation of hares from New South Wales. Later, pure bred greyhounds were introduced. Coursing was banned in New South Wales in 1953. Unless present in large numbers, hares do not do a great deal of damage. In spite of a large heart and the great stamina needed to twist and turn at speed over distances, hares are rather delicate creatures which die easily of nervous shock without suffering actual physical harm. An English north-countryman once described a hare hunted by hounds until it could go no further as 'dying of the black sweat', a grimly effective description of being, literally, run to death.

RABBITS have been the most damaging pests imported into Australia so far. They are voracious eaters but poorly utilise their food in making protein. Eight rabbits will eat as much as a fully grown sheep and produce only a little more than a third of its weight. Rabbits are hardy creatures, adaptable, and highly resistant to shock. One of their most valuable survival capabilities in Australia is that of being able to go for long periods without food and water, with consequent high loss of body weight. They live for up to seven years and reproduce in season at speed. A doe carries her litter for only thirty days before immediately becoming pregnant again. By the time the second litter is born she has raised, and pushed out on to the world, her previous litter. Litters consist of five to ten young and parturition occurs four to six times a year! A mysterious mechanism operates when rabbits become overcrowded locally. Does stop becoming fertile, and pregnant does reverse the order of growth, absorbing the quite well-grown embryos back into the womb walls until nothing of them remains.

Before the rabbit was brought in, there existed in Australia a beautiful small animal called a bilby. The bilby was rabbit-sized, had long ears and a long tail, and fine blue-grey fur. It lived in deep spiral-form burrows, often sharing them with possums and rat-kangaroos. The bilby was in fact a rabbit-bandicoot and was widespread and in large numbers throughout southern Australia. The rabbit appears to have almost completely eliminated the bilby. The last bilby seen in New South Wales was in 1912. A few have been seen elsewhere in recent years.

Five rabbits accompanied the First Fleet to Botany Bay in 1788, and a further importation was made in 1891. Apparently these did not survive. In Tasmania, in 1822, in a climate perhaps approximating better that of Britain, they were reported to be in large numbers and thriving. By 1827 the local newspaper was referring to them 'running about on some large estates by thousands'. There were no known rabbits on the mainland until the 1830s. Within a few years attempts were being made in Victoria to breed and establish rabbits in the hinterland. The fire may have been slow to kindle into flame but the rabbit conflagration that soon followed was to have consequences that are still felt today.

By the 1880s, the rabbits were spilling outward from the host states of South Australia, New South Wales and Victoria, adapting themselves rapidly as they travelled, to link up right across south-eastern Australia. Soon they were living in the arid sandhills of the Centre, the subtropics of the north, and amid the snow and ice of the Snowy Mountains. On the way they learned some of the tricks of survival, broaching the sap of mallee roots in northern Victoria and finding moisture in areas of extremely low rainfall.

An important, but carefully concealed, factor in the spread of the rabbits was their clandestine release on clean land by farmers wanting sport and rabbit trappers with an obvious interest in the rabbits' rapid spread. Human intervention, good conditions and lack of natural enemies turned the rabbits from hundreds into thousands, from thousands into millions, and then into thousands of millions in a furry grey tide that soon completely covered the continent south of the Tropic of Capricorn.

The destruction wrought by the rabbits was almost beyond belief. The Victorian government conceded the loss of 800 000 hectares in the Mallee alone, and in 1887 the rabbits swarmed in South Australia, eating leaves and bark, destroying trees and plants, denuding tree trunks to a height of 47cm from the ground. Drought drove the rabbits to the ultimate destructiveness of eating bark, but although millions perished a few years of good rains had them breeding up beyond their previous numbers. In 1887, the year of the great destruction in South Australia, over 10 million rabbits were destroyed in New South Wales. Estimates of the numbers of rabbits on properties varied but one (said to be an underestimate) for a property of 145 800 hectares came out at 36 million! Fences were inadequate to prevent the migratory movements of the rabbits. It was not unusual for rabbits to overrun the streets and houses and gardens and cellars of towns. They ran in and out of houses and had to be ejected from under beds. By the 1890s the whole of the central west of south-east Australia was infested and over 3 million hectares of land had been let go. Western Australia was the last of the States to receive rabbits. In the year of Federation, 1901, they were firmly established on the goldfields and in the south at Esperance.

Almost every conceivable effort was made to eliminate the rabbits. They were shot, trapped, netted, ferretted, poisoned, gassed, and dug out, fumigated, dynamited, gelignited, petrol-bombed, driven, and finally infected with disease. Poisons, especially strychnine, were used in huge quantities. The method was indiscriminate and the native creatures and birds suffered badly, dying in their thousands. Rabbit poisoning today is a highly specialised craft, with not a little art in it, and involves the use of the poison 1080, sodium fluoracetate, which the rabbits take readily and which is four times as poisonous as cyanide. It is a highly dangerous poison, especially to sheep, dogs, and native creatures and keeps its toxicity out in the open for very long periods. Carrots (or oats in Western Australia) containing 1080 are used as bait, which is distributed by vehicle in settled areas and by air-drop in rough inaccessible country.

The disease myxomatosis originated in Rio de Janeiro in Brazil, where it was first called *rabbit myxoma*, and first came to the notice of Australian agricultural scientists around 1920. Nothing was done until 1924 when a sample of the disease was imported, but no field trials were allowed. Then in 1933 when myxomatosis was anni-hilating Californian rabbits, an Australian, Dr Jean Macnamara, got hold of some of the virus and sent it to a scientist friend in Melbourne. The Director General of Health did not improve of the import and the virus samples were destroyed. After complicated political lobbying, work on the virus was begun by an Australian scientist, Sir Charles Martin, at Cambridge, in 1933. Experiments were carried on in Australia from 1935 until 1943, and then abandoned. Eventually, in 1950, thanks to the persistence of Jean Macnamara, the experiments were reinterpreted and begun again in Victoria. Incredibly, as the disease took hold and spread with unbelievable speed, ferocity, and effectiveness, Jean Macnamara began to get threatening phone calls from people with a vested interest in the fur-and-flesh trade. In a near-hysterical atmosphere the minister in charge of C.S.I.R.O. had to make public the information that three of the country's most famous scientists, Sir MacFarlane Burnet, Dr Ian Clunies Ross, and Professor F. Fenner had inoculated themselves with the virus months before without suffering any ill-effects.

The kill by myxomatosis was over 90 per cent of the rabbit population. By 1953 Australia was almost completely free of the worst of all its pests. The virus was soon being used successfully in both France and Britain. In the year following the virtual annihilation of Australian rabbits the wool clip increased by 32 million kilograms.

Myxomatosis was carried by insects, chiefly mosquitoes, and by the mid 1950s Australia was almost free of rabbits, though resistant animals, which had survived an inoculation of a weak strain of the myxomatosis virus were beginning to appear, and in places to breed up again, particularly in drier areas. Neither the Federal nor the State governments followed up the myxomatosis victory over the rabbits with a concerted drive by all other possible methods against the survivors.

The work with myxomatosis has been carried on in recent years, and some States have continued to infect and release rabbits each year. Virus inoculants have been improved as powders, and freeze-dried, and methods of spreading the new strains that have been developed have been improved. Myxomatosis is still supplemented by the use of the poison 1080, and shooting and trapping, and the rabbit is still a menace, though now a lessened threat to agriculture and to flora and fauna. The great chance of the 1950s was lost, but there always remains the possibility of the development of a strain of the disease which will yet eradicate the rabbit totally. Disease, however repugnant to humanitarians, seems to be the only possible way to a solution. All the other methods, repeated endlessly for almost a hundred years, have proved quite useless.

Before the coming of the white man, with his ignorant enthusiasm, brutal rapacity, and penchant for pests, Australia was the gentlest of all the lands, an authentic, changeless Eden, timeless and calm in its undisturbed unknowing. No great carnivores stalked through the bush and the few men moved across the face of the immemorially familiar land, taking only what they needed, burning with controlled accuracy to maintain the conditions most favourable to their relatively harmless lives in a rare symbiosis between man and nature. The most savage inhabitant to be found, then as now, was man, though the blind barbarism of the white invaders has proved immeasurably greater than that of the Aborigines.

The worst of all the imported pests is not the rabbit; it is the white man. Trail bikes and four-wheel-drive vehicles have made possible damaging sorties into the bush by city savages too lazy to get there otherwise. Their trail is marked by rusting cans and almost indestructible plastic. Erosion follows their churning indifferent wheels, fire follows their carelessness, irreparable damage accompanies their noisy mindless antics. It is possible, if action is not taken soon, that within a few years the vast destruction of the rabbits will be seen as only a minor blemish in a landscape made uninhabitable by the machine-diseased activities of the most destructive pest of all: urban man.

Further Reading

The preceding chapter draws extensively on material in Eric. C. Rolls, *They All Ran Wild*, Angus and Robertson, 1969, a fascinating but badly neglected classic. This book is essential if you're interested.

CHAPTER 6

FOOD

*I look upon it, that he who does not mind his
belly will hardly mind anything else.*

DR JOHNSON

TUCKER is an old English word that had become schoolboy slang
by about the time of the founding of Australia, and thereafter
soon died out in the northern hemisphere. Transported, in who
knows what foul, rotting toothed mouths in the holds of the early
convict ships, it took root in the Australian vocabulary and grew
strongly there forever after, putting out a variety of unexpected
branches such as bush tucker, tucker-bag, gin tucker, tuckerbox, and
even the verb (now rare) *to tucker.*

Bush cooks haven't always been as kindly treated, and often still
aren't. Outback cooks nevertheless throve. They had no competition
on abusive terms like baitlayer, gut starver, belly punisher, poisoner,
crippen (after murderer Dr Harvey Hawley Crippen), and most dryly
Australian of all, water burner. A tradition of gastronomic excellence
is only just beginning to show itself in Australian cities. It has never
shown itself in the bush. Pioneers were not the spoiled and cossetted
people we have become today and they were in contact with food
that we would find hard to take most of the time. When Alfred
Howitt, explorer and anthropologist, first came to Australia he
described his bush tucker like this:

*. . . I have been living on birds — parrots,
parroquets, miners, magpies and quail and gave the
preference decidedly to the first and last. A person can
travel through the country for days with tea, sugar,
some biscuits and a gun.*

It was not long before the settlers and shepherds settled down to
the monotony of mutton, tea and damper. The tea, such as it was —
stations often issued inferior tea full of stalk and tea dust — was
drunk hot, black and sweet. Soda was sometimes added to make it
stronger! Damper as it was made in the 1840s was described by Mrs
Charles Meredith in *Notes and Sketches of New South Wales:*

*. . . I think it the worst possible way of
spoiling flour. The etymology is perhaps 'Dampier', this
indigestible food (an excellent damper of good appetite)
being supposed by some persons to have been invented
by the great circumnavigator, and the manufacture is
this: a stiff dough is made of flour, water, and salt, and*

*kneaded into a large flat cake, two or three inches thick,
and from twelve to eighteen broad. The wood-ashes are
then partially raked from the hot hearth, and the cake
being laid on it, is heaped over with the remaining hot
ashes, and thus bakes. When cut into, it exceeds in
closeness and hard heaviness the worst bread or
pudding I ever tasted, and the outside looks dirty, if it is
not so: still, I have heard many persons conversant with
every comfort and luxury, praise the 'damper'; so I can
only consider my dislike a matter of taste. In 'the bush'
where brewer's yeast cannot be procured, and people are
too idle or ignorant to manufacture a substitute for it
(which is easily done), this indurated dough is the only
kind of bread used, and those who eat it constantly
must have an ostrich's digestion to combat its injurious
effects.*

The formidable Mrs Meredith's severity was unjustified and her
etymology incorrect. Damper does not derive from Dampier, but
from the old English damper, meaning a snack between meals or a
wet blanket, either of which would dampen down fires, whether of
hunger or flame. Bush damper was not made with yeast because it
was unobtainable, but substitutes, including baking powder,
bicarbonate of soda and cream of tartar, and even Eno's Fruit Salts,
though the latter, while superior in the results it achieved, was
usually too expensive for common use. Damper is nourishing but
not particularly palatable and the bushmen gave it suitably
descriptive names: bunghole, nightmare, dorkum, woppidown.
Damper dough can be varied in size and constituents to make small
scones or johnny cake, and brownies or tommies, which are damper
dough enlivened by the addition of fat, sugar and currants. Dough
was also fried in fat in a pan to make 'a flooper in the pan', or made
into small lumps which were dropped into boiling water to make a
soggy variety of dumplings which were called dips.

The ubiquitous mutton accompanied by damper was called
simply the Old Thing. A side of mutton was a concertina, a shoulder
a banjo, and tough meat was called ram-struck mutton. Under-
ground mutton was not mutton at all but rabbit. A boned leg of
mutton with sage, onion and breadcrumb stuffing was called
Colonial Goose, and a stuffed shoulder of mutton was called
Colonial Duck. The only other real meat the bushman might
encounter was kangaroo, and then the tail made into soup was
preferred. Esoteric local dishes included Grabben Gullen pie, which
was pumpkin filled with possum meat, and Gundaroo Bullock which
was koala meat.

Bushmen's names for food were usually expressive of a certain
philosophical cynicism about eating. Custard was called ointment,

honey was bee-jam, jam was fly-bog, cabbage was chow or leprosy, and tinned fish or kippers were called goldfish. One of the many vividly descriptive names applied to golden syrup or treacle was tear-arse.

The best bush recipe is still the one for cooking a galah. This reads: Boil water in a billy. Pluck the galah and place in the boiling water. Add a stone about the size of a fist. When the stone is soft the galah is ready to eat.

MODERN BUSH TUCKER derives from Aboriginal food. In tropical areas where fruit bats or flying foxes are plentiful they are considered to be very good tucker indeed. Wings and membrane are removed close to the body, entrails are left in, and the bodies left to cook amongst hot coals. Cooked in this way the bats will keep for days. This method not only seals in the nutrients but insulates against contamination from outside. When eaten, the skin and flesh is separated from the carcass and entrails, and the meat taken from the skin. Even the bones are soft and edible.

Goannas have their entrails carefully removed but the liver is put back in. The carcass is put on the fire to crisp the scales, then it is transferred to a ground oven.

Snakes need special treatment, having to be heated and stretched over a fire before being prepared for cooking by having incisions made down either side of the back. The entrails are not removed until after cooking, which involves the snake being rolled and tied and put amongst hot coals and ashes. Snake is said to be like chicken in taste, and the fat, as that of the goanna, is highly prized and enjoyed.

Bandicoots are said to be good eating, with flesh that tastes somewhat like rabbit. They can be roasted, baked, and the flesh filleted from the bones and cooked in an ordinary batter.

TRIBAL ABORIGINAL FOOD was inseparable from the turning of the planet, the circle of the seasons, the annually repeated cycles of wind and rain, sun and dryness, heat and cold. There seem usually to have been more than the usual four seasons, all clearly marked and recognised. When certain stars stood in the sky the temperature could be expected to fall and the wind to blow from one direction, making kangaroos scarce but perhaps bringing to the ground ripe seeds from certain plants for grinding, cooking and eating. The seasons that were recognised were local but precisely identified, as for example when, in certain areas, the great heat of high summer arrived it would be time to hunt kangaroos parti-cularly, because their movements in the area would be restricted and their speed reduced. The Aborigines were participants in an endlessly repeated but infinitely varied total life experience in which sun, moon, stars, plants, animals, weather, movements, rituals, arts, and food were inseparable from one another.

Food gathering and hunting were quite leisurely pursuits in areas sparsely peopled. Occasionally there were seasonal feast times when, for instance, the Bogong moths swarmed on the alpine uplands, and the Aborigines journeyed in large numbers from all around to meet and feast on the insects. These occasions of abundance were also an opportunity for ritual practices, and for the trading of articles of particular local value and utility; a certain sort of stone for axes, a special sapling for spear making, ochre for decoration.

Modern research has shown that the predominant food of the Aborigines was vegetarian, that Aboriginal society was more dependent on gathering than hunting, that seeds, fruits and nuts were more important than kangaroos, or reptiles, or fish, crabs and shell-fish. This seems to have been true for much of the continent, though on the coastline fish was usually an important component of diet, and in the cooler south-east, where the keeping of flesh was easier, the kangaroos, fish and birds were of first importance. Little is known of the Aborigines' knowledge of food preserving, though Professor Mulvaney, in *The Prehistory of Australia*, reports 'A resident of the Pilbara, near Roeburne, described both the storage of grain under bark covers and the sun-drying of cooked meat.' Biltong, or smoked, dried raw meat has not been noted as part of Aboriginal diet from any part of Australia.

The division of labour in the gathering of food was with the men of the tribe hunting and fishing and the women gathering and preparing nuts, fruits, eggs, seeds, grubs, plants. While the hunting was the more dramatic activity the results were occasional and limited. Food gathering was the basis of the Aboriginal diet and for most of the time the tribes were dependent on their womenfolk for sustenance, an arrangement that may look sexist enough to modern eyes, but which seemed natural to the Aborigines and persisted unchanged for thousands of years.

Carbohydrates were not missing from the Aboriginal diet. The natives harvested a number of grain seeds, ground them with stones and baked a paste made from them in their fires to make a sort of bread. The grasses were harvested, and the seeds separated by winnowing. In the northern swampland wild rice was harvested. This too was ground up, made into a paste and baked, after being separated from the straw and chaff. The roots of young green gum trees were stripped of their bark, cooked in hot ashes, ground up and eaten. The tuberous yams of the north, and the smaller varieties found as far south as northern New South Wales, were widely used by Aboriginals who cooked them, mashed them, and strained away the inedible fibres and starch before eating the remainder.

A handy practical Aboriginal hint when food is scarce says: to make food go further, eat less.

Food FOR BUSHIES has to be carried along, since sustenance is obtainable from the bush only in emergencies (Chapter 15). The risk of poisoning or near starvation is great if care and thought are not applied to the selection of food, and its cooking. 'A little of what you fancy does you good' is as true in the open air as anywhere else. Food is an ultimate personal matter and most of the long lists of preferred foods in camping books are simply the advertisement of the author's own prejudices. Two things need to be kept in mind. More food will be eaten out of doors than in, because of the hunger following exertion and fresh air, and in selecting foods thought must be taken for the provision of a properly balanced diet which will fuel the body appropriately and adequately, and sometimes more than just adequately when an extra effort has to be made.

Food becomes doubly important in the only intermittently familiar circumstances of the outdoors. Sound nourishment is as important psychologically as physically. Irritation, depression, and inexplicable tiredness can all be caused by an unbalanced diet or badly prepared food. At an extreme, inadequate and unsuitable food can contribute significantly to hypothermia (exposure) which is a lowering of the general body temperature in cold, windy, and wet conditions, which leads to extreme weariness, false euphoria, loss of judgement accompanied by increasing clumsiness followed by collapse, and death (Chapter 15). In some cases only thirty minutes may pass from the initial symptoms to collapse and unconsciousness.

Sound nutrition demands broadly that you have a reasonable daily intake of foods that will keep the body in good repair, provide energy and warmth, and the essential minerals and vitamins which help maintain the body in sound working order in something of the way lubricants do in engines. In practical terms this means that you must have reasonably substantial quantities each day of *one* of the foods in each of these groups: milk, cheese; meat, fish, eggs; fruit, vegetables; butter, margarine. Helpings of these foods will provide all the vitamins, minerals, trace elements, proteins, carbohydrates and calories you will need without having to think any more about them. Only one element is now missing.

The ENJOYMENT OF FOOD is the most important aspect of eating. Keep your four food groups in mind and select what you know from experience you will enjoy most from each group and which you can best afford. Always carry adequate supplies of your food (your known needs plus 10 per cent is a useful formula) and be sure that you have available high energy foods (sugar and glucose sweets, chocolate, Kendal Mint Cake and so on), both to eat at intervals during the day as energy is spent and as a safety pack in case of emergencies which can lead to a sudden big demand for energy. What you eat is your own business as long as the foregoing is kept in mind. It is especially true out-of-doors that one man's meat is another man's poison, and much of the advice offered by books on camping seems to be offered by toothless, tasteless puritans determined to extirpate their appetites by pain and suffering. All food is probably adequate, only the food you thoroughly enjoy is good for you. After your feet, look after your stomach. What follows is personal and highly prejudiced.

A VARIETY OF FOODS specially prepared for outdoor people is now available. Some, like the full cream quick dissolving dried milks, are excellent, being highly portable, nutritious, and flavoursome, as well as being suitable for packing in strong plastic bags that can be moulded to fit into convenient spaces. Others, usually advertised as complete meals, are powdery mixtures, tasting of nothing in particular, weighing nothing much, and are incapable of satisfying any normal appetite. Often their nutritional qualities are extremely dubious even if the hardihood involved in absorbing them is achieved. There are strict limits to what can be done in dehydrating foods. To go below a certain bulk and weight reduces their nutritional value almost to nil. Ultra-lightweight powdered foods are usually ultra-lightweight in content.

There is now available a wide range of continental meats which carry well and can be eaten raw, or cooked, or chopped into soups and stews. Dehydrated fruit (dates, prunes, bananas, apricots, peaches, etc.) and vegetables (peas, carrots, beans, potatoes, etc.) are available in great variety and the fruits can be eaten raw or cooked. Powdered egg is useful, but on shorter trips fresh eggs are better and not very difficult to carry if properly packed. If meat is the King, cheese is the Crown Prince of the house of protein. Properly treated and wrapped in plastic, cheese keeps very well indeed. Oatmeal (fine) is quick to cook (eight to twelve minutes stirring *all* the time) and is infinitely superior to so-called 'quick' oats in substance, nourishment and palatability.

Wholemeal flour is better than white flour and can be used to make delicious Indian chapatis instead of dull, heavy damper. Chapatis are easy to make and best eaten hot, though they *can* be eaten cold when they are slightly tough but tasty and nourishing. Spread with butter they can accompany almost any meal, or be eaten with jam or honey. A cup of wholemeal flour will make four or five chapatis. The flour is combined with water, a little at a time, and a half teaspoonful of salt, to make a loose dough. The dough is kneaded for a few minutes (the more the better) and then left for about an hour when it is kneaded again and a little more water sprinkled on, if necessary. A piece of dough about the size of a golfball or a little larger is broken off and rolled out thin and round like a pancake. A frying pan or metal cooking plate is heated and lightly greased with butter or margarine. The chapati is dropped in the pan and cooked first on one side and then the other at a medium heat. Cooking should be quicker rather than slower. If done properly the chapatis will 'bubble up' all over. Cooking time should be less than a minute for each chapati. Some people add a knob of butter to the dough at the kneading stage.

Rations of food allowed for each day out are a matter for experience and experiment. Needs vary widely from person to person and knowing one's own needs, and catering to one's own tastes, is the most economical way of carrying food in the long run, and better than complying with 'ideal', theoretical lists which may involve carrying unnecessarily a large surplus, or may leave one perpetually hungry and dissatisfied.

Food is important to health and survival. Make sure that a balanced selection that you will *enjoy* is carried with you. The outdoors and the fresh air will improve the taste of even the best things, so catering for oneself isn't at all difficult.

FOR THE BUSH KITCHEN everything except the kitchen sink is a bad formula. Walkabout cooking is necessarily of the simplest sort, particularly in warm weather when prepared food can be carried and when tea is frequently in demand. Basic utensils should enable frying and boiling to be carried on. Baking is best accomplished by using aluminium foil which has many culinary and other uses.

Lightweight cooking sets of strong aluminium are cheap. They are usually made up of a deep billy for boiling, a shallower pan for frying, which doubles as a lid for the billy, a couple of small plates or bowls, a plastic cup (totally inadequate even for midget thirsts), and a detachable handle that fits the cooking utensils. The handle is an important part of the set, simplifying as it does the movement of hot pots to and from the heat. In any set there should be a lid, or a pot that can be used as a lid for the billy. Almost any breeze carries away heat and steeply increases the consumption of precious fuel. A lid helps to prevent this and also keeps out insects, smoke-smuts and ashes. A drawstring-top cotton bag is an essential for carrying cooking pots, keeping them together, and stopping any dirt on them from rubbing off on rucksack and clothing. Cutlery is available in cheap KFS (knife-fork-spoon) sets which clip conveniently together. They are a poor substitute for a good strong stainless steel fork and spoon and a really sharp bladed pocket knife. Strength and durability are important. Spoons in particular have many uses, not the least of which is the digging of a shallow trench for a camp oven. Most foods can be eaten with a spoon and/or knife, but a fork can often save burnt fingers. Aluminium heats up very quickly and it is easy to burn foods, especially when frying, or stewing. An asbestos mat is cheap, light, flat and takes up little space. Under pots it spreads heat evenly and burning becomes almost impossible. MUGS (1) are best made of plastic. Metals get too hot and can give quite nasty lip-burns when not dropped from blistered fingers first of all. Enamel is better but soon chips. Plastic mugs are now available in a number of sizes. MUGS (2) are people who forget to take a can opener with them, and don't have one on their pocket knives.

Beyond these simple essentials, cookery can become as complicated and sophisticated as the transport available; involving portable ovens, pressure cookers, and large heavy gadgets which beat-as-they-sweep-as-they-clean, and don't belong in the peace and quiet of the open air and satisfying basic living.

ALUMINIUM FOIL is one of the handiest creations of modern science. Its uses in the kitchen and for cooking are innumerable. At a pinch it can be folded to make a vessel for boiling water or for cooking. It insulates food well from both heat and cold when used as packaging. If a pot lid goes missing, foil can be used instead. Its best use is in baking. The food to be cooked is wrapped in a double thickness of household foil with edges crimped to make an airtight bag and some space left inside for the expansion of the air while cooking is going on. The bag is then put in, and covered over with, gently glowing ash and embers to cook. Steak smothered in onions and liberally sprinkled with salt and pepper is particularly good cooked in this way. Thinly cut vegetables can be included in the bag and will be cooked through by the time the meat is cooked, in about 20 to 30 minutes. This method is good for almost anything: vegetables, fruit (apples, covered and filled with honey or brown sugar, are especially tasty), fish, chicken, chops, sausages. Household foil should always be used in double thickness, and *never* used in flames. Foil is useful too in keeping vegetables separate when boiling. The different vegetables are enveloped in double foil and a little water and the resultant bags put in a pot of boiling water to cook. Foil makes an excellent fire reflector if you favour this sort of cooking, and can be twisted to make drinking cups if needed. Foil has only one disadvantage: it is virtually indestructible and *must* be carried out with you. Since it compresses easily this is not an onerous chore.

FIRE is probably the most useful, and the most dangerous, of all human discoveries. This is particularly true in Australia, which is the driest of all the continents and the most vulnerable to damage by fire. This is especially so of the bush of the eastern areas. Compared with other lands, Australia is only sparsely forested, and the area (only about 4 per cent of the land mass) is shrinking every year. The land can no longer afford intense destructive bush fires. We must understand that the bush is irreplaceable and exercise the greatest care when using fire in any of its forms.

A FIRE FOR COOKING can usually be easily and quickly made from small dry dead wood lying on the ground, and fed with small dry fallen branches. Good cooking is never possible on a blazing fire and to get the bed of glowing, smokeless ash which is the best and most easily managed for any sort of cooking, small dead branches are quickest. For cooking the aim is the creation not of a fire, but of stove heat which also keeps pots much cleaner than the soot and smoke of flaming fire. There is no need to carry toma-hawks, axes, or bow saws unless the end in sight is a bonfire beacon, which is a dangerous thing to make in the bush anyway, unless in an emergency. Fires for warmth should be built only when absolutely necessary, and green growing wood should *never* be used, not even for small cooking fires. On Fire Ban days it is forbidden to light any sort of fire at all, even in an emergency. One person's emergency could too easily become a community's catastrophe.

A fireplace of stones or logs helps control the fire, ensures good draughts, and can be used to support pots. Clear the ground for a metre around it. Start the fire by laying the smallest, thinnest, and therefore most easily combustible, material in the fireplace. Have stacked ready some larger twigs broken into suitable lengths, and beside that some larger material again. When the small wood is well alight, feed in the larger twigs; when they are blazing build up the fire with the larger, thicker, branch material. No cooking fire needs sticks more than a couple of centimetres thick. Until this last material is burning satisfactorily give all your attention to the fire,

feeding it when necessary, blowing on it if you think it needs it, and generally coaxing it along. The way you lay your material is up to you. Some people arrange it in teepee shape, some in criss-cross grids, some seem to just drop it as it comes. The way it works for you is the best way. Some fuel is better than others, but experience is the best guide in selecting your firewood.

DAMP AND WET can make firelighting difficult. When starting a fire at any time it is best to get your kindling, the thinnest of dead twigs, from bushes. Small bits on the ground will almost certainly be damp and slow to burn unless there has been a long dry period. If the ground is wet, lay your fire in a protected spot on a double layer of foil. Dry kindling can be found under logs and under thick bushes, and trees; ferns and shavings can be whittled. Fuel can be got from dead branches of standing timber and wet outer bark removed with a pocket or sheath knife. The dry underside of the bark of the stringybark tree can be scraped and teased out to make very good kindling. If your pack is properly packed you will have in it some old candle ends, solid fuel tablets, or a strip or two of car or bicycle inner tube rubber, all of which burn well and help to get a fire going even in very wet conditions. Once the fire is going and enough heat is generated, wet fuels can be dried and covered for use later. There has also come on the market recently a very efficient inflammable jelly which can be squeezed from a tube or taken in a gob from a small tin to help start a fire.

MATCHES are still the best way of lighting fires. Lighters break down and life is too short for rubbing sticks together. Damp-resistant matches (such as Greenlites) are best for general use, but even these perform better if kept dry in a screw-topped container or wrapped in a small heat-sealed plastic bag. Carry several boxes of matches distributed through your pockets and your rucksack.

CARELESSNESS COSTS LIVES and can destroy irreplaceable bush. Select your fire site with care, well away from inflammable materials like fallen logs and stumps. Take extra care in hot, dry weather. When you move on, make sure that your fire is *completely* out. Pour plenty of water all around the site as well as on the fire until the area is soaked. If no water is available, cover to a depth of several centimetres with soil. There is *always* risk with fire and there is a good argument that says it is best to use a small stove as much as possible and make use of fire only in emergency or when the carrying of fuel in sufficient quantities for the outing is made difficult by the sheer volume required.

COOKING STOVES begin with collapsible bits of tin which, when put together, make a holder for chemical solid fuel tablets on which a pot may be rested. Some of these tin stoves come with an equally collapsible windshield. Solid fuel is an inadequate heat producer for most cooking needs except the simplest and the heat produced cannot be regulated. The Russians may have reached outer space with solid fuel rockets but their chemical expertise has not yet got through to the preparation of food for inner space. Fuel tablets are still useful, light and cheap, for getting a fire going in wet country.

LIQUID FUELS include petrol and kerosene. The petrol is a specially prepared clean fuel, quite different from the stuff in petrol bowsers which should never be used in cooking stoves. A number of petrol stoves of different design are available. All work well but require extra care in their handling. Petrol is a highly volatile fuel. Petrol vapour ignites and expands with explosive rapidity. These stoves should always be handled with the greatest possible respect. The heat from petrol stoves can, and must, be regulated properly, lest the safety valve come into operation like a wartime flame thrower. Petrol stoves are usually crabbed, demanding, and temperamental things with a special appeal for the eccentric and the idiosyncratic, who manage them well and will use no other.

Kerosene is probably the best of liquid fuels, but it too has disadvantages which have to be kept in mind. The chief of these is its smell, which will unpleasantly permeate everything within reach, including food, if not adequately restrained in stove and container.

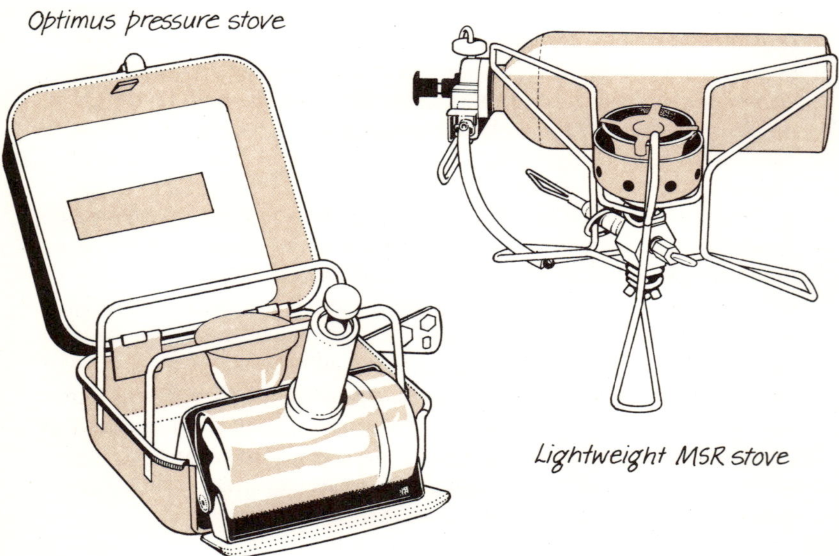

Optimus pressure stove

Lightweight MSR stove

Gas burner, gas stove and cooking utensils.

The kerosene stoves need a supply of methylated spirit, as well as fuel for initial warming and ignition, and are 'pumped up' to give burner pressure. Although not as hot as petrol cookers, they are safer and rather less temperamental; though anyone who has wrestled with one in intense cold, with meths evaporating refrigeratingly from the fingers and a bent pricker (for clearing the jet) refusing stubbornly to do its job, may feel that this is too generous an assessment of their capabilities.

With both petrol and kerosene stoves, especially if using thin aluminium utensils, it is a great help to cooking to carry also a small asbestos mat. This spreads heat evenly and gently, preventing burning and food sticking to pans but should obviously not be used when the quickest maximum heat (boiling water) is needed.

Methylated spirit stoves are now seldom used, except on yachts where there is the special problem of dangerous accumulation of vapours and gases in the bilges and the consequent risk of explosion. The heat generated by methylated spirit stoves is much lower than that of petrol or kerosene so cooking is much slower. Where greater heat is generated much more fuel has to be carried.

G AS, usually butane, is now the most common of all the fuels used by campers, caravanners, and bushwalkers. Gas stoves come in all sizes and so do the gas containers. In small, single burner, lightweight stoves there is usually a pot stand and valve element combined, to which is plugged a gas container cartridge. Although rather less efficient in heat production, gas stoves are safe, dependable, quick-starting, and very clean to use. The several disadvantages are that the fuel is rather heavy, efficiency is impaired once the temperature drops below freezing, the empty cartridges are *not* empty but still capable of exploding, and they should be carried home for normal garbage disposal. For short trips in warm or temperate weather where weight is not an important consideration this sort of stove is almost unbeatable. Stoves are made in a number of models for use with gas. At one end of the range is the standing four-burner with oven stove for caravans and country domestic use, and at the other the almost pocketable miniature stove for the bushwalker, with the plug-in cartridge.

O UTDOOR COOKING, whether on fire or stove, is largely a matter of commonsense. Any flame burns with a better heat if protected from the wind. Stoves, especially the taller variety, are more stable if they are chocked round with stones or pressed firmly down in a shallow hollow of sand or earth. Fires need a fireplace of some sort. Many stoves have a burner control to regulate the flame. Fires are regulated by altering the distance of the cooking pot from the bed of glowing embers, either by building the fireplace up a stone or two or suspending the pot from a dead branch set in a forked twig and firmly held down at one end. You will need wire for making hooks if this is done.

B USH OVENS are easily made by digging out a small trench a few centimetres deep and lighting a fire in it. When the fire has burned down the ashes are brushed aside and heated flat stones used to line the bottom. (Don't use stones from creek or river beds, they are likely to explode when heated.) The food, wrapped in a double layer of foil (or tea-tree bark if no foil is available), is put on the hot stones and buried in the hot ashes. A medium fire is then lighted on top and allowed to continue burning until the food is cooked. This oven is particularly good for joints or large pieces of meat.

W ATER is never more useful than when it is in short supply, or obtainable only in small vessels. A collapsible lightweight water bucket weighs only a few grams and stores away easily in a small space. It can save many small journeys and inconveniences at the end of a tiring day, and allows a better washing up to be done. Pots and pans are best cleaned by rubbing with sand or soil and then washing in water, but there are advocates of small cleaning pads of wire wool impregnated with soap. One pad can be used repeatedly, occupies almost no space and is hygienically superior. Pots, pans and cutlery should always be cleaned after use. Dirty utensils are not only unpleasant to use, they introduce the possibility of stomach ailments when usually far from any medical help.

CLEANING UP the cooking site when leaving is always an important part of conservation. Rubbish should be flattened and carried out, together with any plastic or foil. A stout plastic rubbish bag will take care of this problem and the weight is negligible. Burn paper and anything else that can be reduced to ash. Scatter ashes, and leave the cooking area as much like it was on arrival as is humanly possible. The old rule of burn, bash and bury has been made obsolete by a mechanical society intent on suicide through the destruction of its natural resources. The bush is shrinking fast and its use will soon have to be rigidly controlled unless bushwalkers see to their responsibilities in this way. Already there are reports from many parts of the world of the wearing out and massive pollution of nature trails by the sheer weight of numbers of the people using them. In Australia this has only recently become a problem in bush areas close to cities. *But it has become a problem and a growing one.* The bush is much smaller than was once thought and everything that can be done to preserve it, no matter how small or seemingly unimportant, must be done.

Further Reading

Most books on camping have a chapter on food according to the prejudices of the author. This one has.
L. Stirling, *Cooking in the Bush,* Albatross, 1984.
Wessa and Lummo, *Bush Cooking,* Rabbit on a Spade, 1988.

PRACTICAL CAMPING

Down in the jungle
Living in a tent
Better than a prefab —
No rent!

CHILDREN'S RHYME

ULTRA LIGHTWEIGHT CAMPING is the most easily practised form because it requires the minimum of equipment; only one article needs to be carried. Unfortunately it is really an overnight form demanding ingenuity and involving a little discomfort. Carried on over several days it demands some hardihood. It is also relatively expensive. Normal clothing and a pocketable plastic mac are all that are needed except for the one article of camping equipment which also slips into a pocket. When night falls shelter is sought in a haystack, on a bed of grass or reeds or brush, or under a thick cover of bush. Clothes are kept on and the plastic mac used as extra cover if wet weather develops. A good, comfortable night's sleep is assured by nipping frequently at the half-bottle of rum (or brandy, or whisky) which is the only article of camping equipment carried.

LIGHTWEIGHT CAMPING is probably the most satisfying of all because it gives a wonderful feeling of total independence and containment if done properly, which means having the right sort of equipment in a complete outfit. Expert lightweight campers begin their excursions in the bathroom with the pack on the bathroom scales, having previously found out by experiment the most comfortable load they are capable of carrying for any distance. Expert lightweight campers are ruthless people. Having found their most comfortable carrying weight, they will not carry a gram more. If the pack is overweight they spread out the contents to see what they can best do without and then leave the article or articles at home. Naturally, they leave nothing behind that might ensure their safety and survival in an emergency.

Lightweight campers are usually bush or mountain *walkers,* though the same equipment, in general, is used by canoeists and cyclists who still are a small minority in comparison. All are found with the same sorts of camping problems in their daily travels. Their nightly need is palatable food, relaxation, and comfortable rest. Most time and energy are spent in movement during the day and a routine for stopping and making camp saves time, energy and temper each evening. A good routine includes previous agreement on minimum campsite features. This is important because tired people will accept unsuitable, and possibly dangerous, campsites, at which, apart from the risks involved, they will probably pass an uncomfortable semi-sleepless night, and set off next day already at the beginning of a self-generated vicious circle of fatigue leading to bad judgement, fatigue, and so on. Camping is essentially for enjoyment and tired, bad-tempered people don't enjoy it and can quickly become a risk and a nuisance to their companions.

M INIMUM CAMPSITE FEATURES are simple. Unless water (which is very heavy stuff) is carried, the campsite should be close to a supply of fresh, *fast-running* water. The site should be in an adequately sized clearing so that tents can be pitched well away from trees (and the risk of blown or fallen branches). A level, firm site, away from damp or marshy ground is best, and plenty of dead wood around helps to conserve cooker fuel supplies. Flies, sandflies and mosquitoes are usually unavoidable, but the campsite should be carefully examined for the presence of ants or other insect pests, especially if you are settling in at dusk. Taking down carefully pitched tents and moving camp in a dark and unknown area when tired after a long day's walk is a nasty chore. Equally unpleasant is to wake up to the discomfort of a sopping sleeping bag and the lap-lap of rising water, because the tent is pitched beside a rapidly rising stream, or in a hollow, or across a natural drainage depression. Travel should always stop well before dusk so that there is plenty of light left by which to select a good campsite, pitch tents, and make a meal.

E FFICIENT TENT PITCHING is largely a matter of practice at home in the garden, until it is a routine which can be carried out quickly and efficiently in all weather and conditions. If the tent is new, practise pitching it in the back garden and try to have at least one night's sleep in it, preferably in bad weather, before setting out. Tents are individual and the more experience one has of them before use in the remote wilds, the better. A tent flysheet is an almost indispensable adjunct, cooling the tent in heat and protecting it against the worst wind and rain. Practice should enable the tent to be pitched neatly and quickly. If the tent has guy lines running to pegs, be sure to slacken them off a little before going to sleep. Absolute beginners should remember that the correct way to start on pitching a lightweight tent is by first pegging it out on the ground at the corners, and that tent pegs are not designed to go into the ground vertically, but at an angle of about forty-five degrees. They should also remember that backyard lawns are the only pieces of ground campers are ever likely to find where tent pegs slide in smoothly without encountering awkwardly placed stones or other obstacles, and that guy ropes can be attached to, and held down by, things other than tent pegs.

FIRE remains the ever-present danger in Australian bush-camping, and the most stringent precautions always need to be taken. Cooking fires of dead wood and the minimum size consistent with efficiency should always be restrained and contained in a stone or thick log fireplace in the centre of an area cleared of dry leaves and twigs. Gather kindling for the breakfast cooking fire in the morning and stow it just inside a tent to keep dry until needed. It is sensible to site the fire so that smoke does not blow into the tents or into the cook's face, and so that any sparks disperse harmlessly. Dry grass ignites especially easily. Cooking stoves should be handled with equal care and preferably be chocked firmly in place by stones. A roaring campfire of dead wood was once a wonderful source of warmth and light on chilly nights. Such is the mounting demand on bush and wilderness that this is now a luxury that can no longer be afforded, unless life is actually at risk. Every large campfire lighted now is another diminution of our fast-disappearing bush. Burn dead material actually on the ground and then only when it is essential for cooking. On some campsites in Californian parks, all the dead wood has gone, and wood has to be *purchased* from the local ranger. This situation is now not so far off in Australia. *Never cut down any standing timber* whether it appears alive or dead.

A BIT OF QUAINTNESS we can no longer afford is the 'backwoodsman' olde worlde woodsmanship, practised for so long by youth organisations like the Scouts, and by frustrated lumberjacks and 'bushmen' of the stamp of the early pioneers who thoughtlessly assumed the endless availability of bush timber. Axemanship, of a very rudimentary kind, is still implicitly applauded in a number of Scout Association and other publications on sale in Australia. One American Scout publication on sale here about camping has illustrated instructions about how to chop down a tree; but no distinction is made between live and dead trees! Other 'bush' publications include instruction for making thatched huts of various sorts, and camp 'furniture' including bed frames, chairs, tables, hammocks, ladders, and a highly improbable coat-hanger on which, presumably, tail coats may be safely hung. This sort of irresponsibility is no longer as comical as it once seemed, and with the daily onslaughts now being made on the bush, is no longer even a possibility to be contemplated.

BREAKING AN OVERNIGHT CAMP needs a routine just as much as making one does. Tents and sleeping bags *need* a short airing after overnight occupation and are probably best left till last to be packed. Cooking fires should be completely doused with water or smothered thickly with earth after combustible rubbish has been burned. Other rubbish should be flattened or compressed *and carried out*. Dishwashing should be done in a cooking pot and *not* in the nearby stream. Soap and detergents aren't really necessary so long as some sandy soil is available (and it usually is) for scouring pots, plates, and cutlery. Get rid of washing wastes, if you *have* to use soap or detergent, in a bit of bush where the least harm will be done by them. *Never* get rid of wastes into running water where contamination will occur and the health of other people far downstream from you may be endangered. Twenty years ago it used to be said that contamination cleared itself in running water within 10 metres of the source. This is no longer true anywhere in the world, and even running water has to be treated with suspicion. All sources of water should be examined well upstream, for at least 100 metres for sources of pollution. This applies even in the remotest bush where dead animals can be busily decomposing in the water just round the first stream bend.

Toilet leavings (well away from campsite) should be covered with soil in order not to encourage flies. If the campsite is a regular one and has a fireplace already built, leave it. Use existing cooking fire sites whenever and wherever possible. To make a new fireplace when one already exists is wantonly wasteful and damaging. If the campsite is not regularly used return the fireplace stones to where you got them and remove, as far as possible, all traces of the fire and of your brief occupation. If there has been heavy rain overnight and the tent is wet, shake off as much moisture as possible before lashing it around the outside of a pack. As soon as possible (lunch stop if it's warm and dry) spread it out and get it dry as quickly as possible. Don't *ever* try to wring moisture out of a wet tent. It will ruin it.

A STANDING CAMP which may continue in existence for many days, serving as a base from which the surrounding country may be explored, or journeys into remote wilderness areas carried out demands a great deal of efficient planning and practical organisation, but the basic principles remain the same. A very much larger site is required, but water, wood, protected openness and suitable ground (level, well-drained, not subject to flooding) on which the tent may be pitched are still the basic requirements.

Standing camps need to be chosen so that they can be reached by motor transport. Large tents, equipment, and big quantities of supplies weigh heavily and need to be delivered to the site. With motor transport which can carry lightweight collapsible beds, chairs, tables and other amenities there is no excuse for hacking the bush about in misplaced displays of 'axemanship' in order to create, at a cost beyond estimation, inferior and less efficient and comfortable domestic articles.

Commonsense planning, in plenty of time, with a preliminary reconnaissance to establish the location of the campsite, will ensure a camp easy to set up, run, and dismantle. Space is important between tents and cooking area. Even with plenty of space between them, thick smoke drifting through the camp whenever a meal is cooked is an irritation and a discomfort. Toilets need to be deeply dug, out of sight and where contamination of the water and food supplies is impossible. A downwind location clears any smell away, but if a thin layer of the earth dug out to make the toilet is thrown in after each use so that leavings are completely covered, this should not be a problem. Earth heaps should each have a small shovel or garden trowel left stuck in them. It is usually worthwhile to build a quite large fireplace if one does not already exist. Cooking for a number of people is a complicated business and anything to make the 'bait-layer's' job easier should be done.

The function of the camp should be kept clearly in mind. If it is a purely recreational holiday camp, extra space for activities will be needed, and bush tracks and a river or lake nearby will be valuable assets. A base camp enabling walkers to explore, climbers to climb, or difficult country to be reached safely has to function differently, providing rest and relaxation, probably acting as a source of stores and equipment, and having quite substantial medical and first-aid resources, as well as rescue equipment. A good base camp is also a place where communication with more urbanised areas and their technological resources (helicopter, four-wheel-drive vehicles, search planes) is quick and easy if an emergency of any sort should arise. Swift transport to the nearest telephone should always be kept available in reserve.

Standing camps accumulate rubbish at a prodigious rate. A pit should be dug solely for liquids like washing up water, saucepan gravy, and goo of all sorts. Paper, with other combustibles, should be burned *daily,* not left to a cleaning squad on the last day after having made the camp look like a rural slum in between times. Pits have to be filled in before leaving and indestructibles (foil, cans, bottles, etc.) *must be taken out.* With transport this is easily done, and failure is inexcusable. If any rubbish already there in the bush on arrival can be taken out on leaving, so much the better, for us, and the bush. We can no longer tolerate the concept of the bush as a useless wasteland and national rubbish dump.

HUTS are a feature of the high plains and hills of the Australian alps and are fairly common in the mountainous areas of Victoria, New South Wales, and the Australian Capital Territory. These rough shelters (some are rougher than others) were mostly put up originally by the cattlemen who had beasts grazing the high slopes. In skiing and climbing areas these have been supplemented by huts put up by interested clubs. The majority of huts are left open for use by walkers and are an important aid to safety in the mountains where weather can change catastrophically even from hour to hour and a substantial refuge may become necessary, for a chilled party nearing exhaustion, at almost a moment's notice.

Walkers may use open huts to sleep in overnight and this makes a pleasant holiday from site selecting and tent pitching, and the chores involved in striking camp next morning. However, sleeping bag and bedding and cooking equipment, with the other gear normally carried, are still necessary. Even if planning to use huts it is best to carry tent and groundsheet all the way because popular huts may be found to be fully occupied.

It is important to realise that most of these huts are *basic shelters* and little more, and walkers must take in everything they need. Despite their basic nature, huts should be treated with respect and care. If they contain firewood, it should be replaced before leaving, and the hut left neat, tidy and inhabitable. Walkers are usually sensible people with an understanding, grown from experience, of what a hut can mean to a party benighted in snow or placed in hazard by a sudden savage change of weather. Over the years there have been a few instances of senseless, destructive vandalism and this has led to a few huts being closed. Luckily, the instances have been few, but as more people go to the mountains the risk increases and the need for education grows. Without the huts the mountains would be far more dangerous places than they are. Huts can save lives, perhaps yours one day. Leave them in good order.

T HE YOUTH HOSTELS ASSOCIATION (Y.H.A.). In Britain and Europe Y.H.A. aims at providing simple dormitory accommodation and cooking facilities in hostels that are set about a day's walk apart. The movement developed later in Australia and has grown up with some interesting variations, especially in Victoria, where it not only provided some hostels for 'young people of all ages' but pioneered a programme of outdoor activities which has expanded to include sailing, scuba diving, ski touring, natural history study, and orienteering. The Y.H.A. is open to both sexes and all ages and its hostels are a useful form of camping, particularly for young people travelling with light pockets and the minimum of equipment. To use hostels exclusively, the only essential is a sheet sleeping bag inner for use with the blankets provided at the association's hostels. The annual A.Y.H.A. subscription is only a small one, overnight charges are minimal, and the hostels are a good way to begin a simple form of camping in company. Hostels are made possible by co-operation. Members using them overnight are expected to perform some simple duty such as chopping firewood, folding bunkroom blankets, or sweeping out rooms. Without these small chores the cost of staffing and maintenance would be prohibitive.

In a country the size of Australia, it is seldom possible to have hostels only a day's walk apart. In this way they differ from British and European hostels, being used more as bases for local exploration. The only exception is Tasmania where it is possible to see the whole of the island staying overnight at hostels. In comparison with overseas countries, Australia is only thinly served with hostels, having in all only about a hundred. There are none in the Northern Territory, and the only hostel in the Australian Capital Territory is in Canberra and is run by the New South Wales section of the A.Y.H.A.

The Victorian Y.H.A. runs an interesting variation of hostelling. This is a 'portable hostel' scheme of mobile canvas camps which are operated in public camping grounds around the State, particularly at holiday times. The person in charge of the camp is appointed by a Y.H.A. committee. The 'portable hostel' is made up of five eight-man tents in which camp stretchers are provided. Campers are expected to bring their own sleeping bags and share in the chores and duties. Food preparation is in a portable kitchen powered by bottled gas, and food is bought collectively. The camps are usually in areas with good sporting facilities and abundant sources of local interest.

CANOEING, CYCLING, AND PONY-TREKKING CAMPING are all lightweight forms of camping, and present the same problems as those of the walker. Cyclists carry their equipment in panniers on either side of the rear, and sometimes front, wheels as well as on a carrier rack above the rear mudguard. Canoeists are restricted by the space available in the type of canoe they favour, and pony trekkers have a slight, but small advantage in that a packhorse usually travels with the group, but has to carry the gear belonging to a number of people.

Canoeists are best off for the choice of camping sites along the banks of rivers where few people, if any, have been before them. They are also the most weather-threatened. A sudden rainstorm many kilometres away can set their river rising and roaring into a flood with only small warning. Special care needs to be taken therefore in selecting a camping site so that it is either well above sudden flood level or so that any unexpected rise in water level is immediately obvious. Pony trekkers, too, are able to penetrate into remote and unspoiled areas utilising the old cattle droving trails. The usual requirements for a campsite apply, but in addition good grazing for the horses is needed and an adequate supply of drinking water.

Cyclists are the most restricted in their choice of campsite. Pedal cycles can only be used on bitumen roads or really good bush roads. To attempt to use them on rough bush roads or walking tracks is to ask for trouble from punctures, twigs in wheel spokes, and debris dangerously fouling chains and brakes.

PRACTICAL CAMPING is mostly a matter of practical commonsense and a little forethought. Sound decisions about campsites cannot be made in a fuddled state of tiredness. Rest, relax and eat a snack before finally deciding whether where you are is where you want to stay overnight.

Further Reading

R. Edwards, *Skills of the Australian Bushman*, Rigby, 1979.
R. Greaves, *Australian Bushcraft*, Dymocks, 1984.
Paddy Pallin's Bushwalking and Camping, Sydney, 1985.
Chapman and Chapman, *Bushwalking in Australia*, Lonely Planet Books, 1988.

CHAPTER 8

NATURAL COVER

Already the stream side is dotted with
clusters of upadiddle and old man's foot and
the curious may find in crannies of old walls
the lovely bedsoxia, with its trailing stamen
and its inverted corolla.

BEACHCOMBER

A planet cooling out of the incandescent fires of creation is bare rock. Around it, out of the heat and chaos, clouds condense, seas form, and winds generated by planetary motion begin their breathing. Soon there is rain and sleet and snowfall and the infinite variation of cloud patterning against the blue-seeming blackness of universal space. Atmospheric pressure rises and falls ceaselessly as the atmosphere slides around the surface of the globe, dragged and dragooned by the swift, irresistible movement of matter. The planet circles the sun, adding another movement to its perturbations; temperatures rise and fall rhythmically, each day, each night, each year, and daily weather and seasonal climate are born.

Time is the other dimension of weather, and given enough of it there must be life. Time must pass, slow time beyond the conceiving of human consciousness, while the water wears away the rock, the stones crack and break in the heat and cold of the primaeval seasons, and the winds snatch and abrade the planetary surface with its own fragments, until a rich skin of dust accumulates that deepens slowly into soil, in which the vegetation on which animate life itself will be dependent can grow and begin its evolution.

The first forms of plant life were probably the simplest: bacteria, spores, algae, lichens, mosses. Bacteria are known to have existed some 2400 million years ago. The simplest plants were in existence about 400 million years ago, and it was from this common stock that the rest developed. Much later came ferns, and then pine trees, and then about 100 million years ago flowering plants developed which spread quickly across the planet. In the spacious unlimitedness of geological time there was a gradual diversification. Most of the plants we know today, both flowering and non-flowering, became established about 70 million years ago, and thereafter developed variations to cope with their location and climate. The local conditions often brought a surprising uniformity, a stock of common characteristics, to plants of very different origins.

AUSTRALIAN PLANT LIFE took a pathway to separate development when the land mass became separated from the single original great continent of the earth's surface. From then on it could interact only with itself, and with the very occasional plant growing from seed that was brought from other lands on storm debris, or caught in the feathering of some accidental migrant bird, or blown safely across great distances at high altitudes. Isolation from the flow of the rest of the world's plant evolution began early and only ended with the coming of the First Fleet.

Amongst the oldest tree inhabitants is the beech, which is believed to have come here while Australia was still a part of the larger land mass. At this end of their long time in Australia the beeches have dwindled to two varieties in Tasmania and two occurrences on the mainland in northern New South Wales and southern Queensland. The Tasmanian shrub-beech is conspicuous because of its bright green leaves. The trees on the Macpherson Range on the mainland grow to a great age; somewhere around a thousand years.

As old as the beech, if not older, is the pine, whose forests are believed to have covered much of the primaeval world. There are numbers of sorts of pine on the mainland and again in Tasmania which has both the celery-top pine, the tall King William pine, and the now rare Huon pine, which is unequalled in its qualities by any other wood, especially for shipbuilding. The commonest pines on the mainland are the cypress pines which occur in great variety through the eastern States. The white cyprus pine, of which there are forests in Victoria and New South Wales, is famous for its white, ant-resistant timber. Pines are believed to go back some 300 million years in the pre-history of the continent.

Later, and most probably from the tropical north, came the flowering plants. Spreading quickly to the edges of arid areas they soon adapted themselves to the dry soil and meagre rainfall and established themselves vigorously across the land, becoming xerophytes or dry-livers in the process. Alternations of climate, moist and warm, and hot and dry, encouraged the growth in Australia of both jungle and xerophytes.

The plants of dry areas are water hoarders. They are the most widespread, common and typically Australian plants. Amongst many species some of the best known are the casuarinas, the eucalypts, the tea-trees and the banksias. Water hoarders have thick bark, very long roots to take up whatever moisture is available, and tough or narrow small-pored leaves. They flourish in most areas and grow successfully on the inadequate poverty-stricken soil of sandhills, in stony areas, and in shallow infertile soils. Their seeds keep their life for long periods of time and heat. About 12 000 varieties of Australian flowering plants are now known. Dry plants cover an immense range, from tiny wild flowers of exquisite small beauty, like Sturt's desert pea, to the saltbush and spinifex of the desert areas and distinctive mallee scrub, to the giant mountain ash of Victoria up to 95 metres high.

The eucalypts which are of the myrtle family are the most common, and have been estimated to include 75 per cent of, Australian trees. They are genuine natives and there are over five hundred species, over a hundred of which are now grown extensively overseas. Australian eucalypts are highly valued abroad for their high quality hardwood, quick growth, and their potential as ornamental trees. The greatest eucalypts, amongst the tallest trees in the world, grow in forests in southern areas of the continent. In the south of Western Australia there are the giant karri trees, and 3000 kilometres away in Gippsland in forests covering land up to 1800 metres in height, the giant mountain ash, the alpine ash and the brown stringybark grow. The tall forests continue in Tasmania, often on high hill slopes. Lesser forests cover much of the east coast and extend far inland to the west.

Eucalypts give the bush its distinctively fragrant smells. The leaves and stems are rich in oil, of which 'eucalyptus oil' is the most common and characteristic. Few eucalypts are unpleasant smelling, and one species which grows in New South Wales has an oil with a perfume like that of roses. Amongst the gums are peppermint and lemon-scented trees; their odours best enjoyed by crushing a leaf in the hand. Gum-barked eucalypts are easily identifiable because of their smooth bark which they shed periodically in peeling strips or flakes. A few gums (swamp, manna) have a permanent bark, but this only continues part way up the trunk. Other eucalypts are characterised as boxbark, stringybark, and ironbark.

Eucalypts are naturally fire-resistant. This is because of the high insulating quality of the outer bark which can be badly charred without killing the tree. Of all the varieties of trees on the planet, the eucalypts are the most resistant to fire. While the bark protects the wood, the wood itself has to be subjected to continuous intense heat for a quite long period before it will burst into flame. After low-intensity fires the eucalypts are able to regenerate themselves with astonishing speed. Only closely recurrent fires or fires of great intensity can kill eucalypts.

The brief, controlled, occasional burning by the Aborigines, before the coming of the white man, kept down excessive undergrowth and kept the forest trees growing healthily. These low-intensity fires also took care of the dead leaves and debris fallen from the eucalypts before they could accumulate sufficiently to build up into a big, high-intensity bushfire of the sort that has occurred quite regularly every few years since European occupation. Fire of low intensity plays an important part in maintaining the balance of the eucalypt forests.

The tree most susceptible to this danger of accumulated leaves and debris is the mountain ash. This tree can pile up so much of its own debris around its base that when fire occurs, the bark, despite its high insulating qualities, is burned through all round and the tree dies.

Almost a national symbol, the wattle, which is a member of the *Acacia* genus, is the next most important constituent of the bush after the eucalypt. Beyond the bush, omnipotent in desert and semi-desert, spreads another acacia, the grey-green bush called mulga. There are more varieties of acacia than there are of eucalypt, and most of them are Australian natives distinct from other varieties found throughout the world.

THE AUSTRALIAN JUNGLE which is really, and more accurately, tropical rainforest, is an intruder from the other tropical zones of the world, and the least typical ground cover of the continent. The plants are similar, if not always identical with, other tropical areas, and the chief difference is in the animals who live in it. Giant snails, tree-climbing kangaroos, and giant rodents are among the exclusively Australian inhabitants. The plants are often aggressive, unfriendly and armed with razor-sharp thorns and the capacity to sting passers-by painfully. A macabre parasite is the fig that takes up residence in a home tree and then slowly strangles it to death. Fungi spread sinisterly everywhere, gorging themselves on the heady decay of rotting plants.

The jungle which was once round much of the north-east of Australia, luxuriating on the rich decomposed basalt soil, has nearly gone. Great fields of sugar-cane now march northwards for hundreds of kilometres where once jungle proliferated. Agriculture has supplanted jungle and with it has come the swift brief flare of stands of cane being burnt off, and the sickly sweet unpleasant smell of singed sugar. Jungle can only be found today in reserves or in remote and inaccessible areas inland. In the rush to destroy the jungle and exploit the soil many varieties of rare and valuable timber were almost completely obliterated.

All of the Australian forests have been constantly under savage attack since the first settlement. Greed, stupidity, and the apparent abundance of trees led to the most futile and destructive of uniquely Australian practices: large-scale ring-barking. Superficially, the result of the death of the trees was an improvement of the native grasses for grazing without the bother of removing the trees. Subsequently and inevitably there followed erosion and denudation. Many eucalypts sent out suckers from the roots and growth began again, requiring regular clearing. The ultimate result of ring-barking was not improved agriculture but the destruction of irreplaceable resources and the creation of desolation.

The destruction of the red cedar must serve as an example of the many fine woods destroyed by the first pioneers. Magnificent cedar trees grew from an area south of Sydney right up into Queensland. The first cedar was exported to India in 1775. By the beginning of the twentieth century most of the huge stands of trees, which grew up to 30 metres in height and up to 2.5 metres in diameter, were utterly destroyed. The waste was prodigious. Cedar was not only exported, but used for crude building locally, its richness concealed by paint and whitewash! Much was felled and left to rot.

Not surprisingly, Australia is now an *importer* of timber and timber products. Softwoods are being grown widely in the eastern States in an attempt to reduce the annual import bill. A new threat is the utilisation of the eucalypts of the dry areas in the manufacture of chipboard for export. Conservationists are trying to restrict the spread of this new industry, not always with success. If continued there is bound to be irreversible change in much forest land that is most typically Australian. The semi-desert of the interior is already becoming true desert, and it often seems that the politicians and 'practical' men of present-day Australia will not be satisfied until the whole country, from coast to coast, is another waste of destruction.

Near ground level there is an unbelievable abundance of small plant life so rich that many of the varieties growing in quite small local areas have yet to be identified and recorded. The Australian wild flowers, like the native trees, have adapted themselves to every extreme of climate, from intense heat to alpine coldness, and from dry aridity to all-pervasive moisture and to all the combinations and permutations that these can form in relationships with a variety of soils.

The hardiest of these plants live high up in the small alpine region of the Snowy Mountains. Here, exposed to freezing winds, snow, sleet and rain, and living out of an always cold, damp soil, are plants which survive by living flattened closely to the ground in thick mats holding co-operatively together to defeat the worst that an adverse climate and environment can do. The winter snows cover them deeply but they survive into spring and summer in their thickly interlaced small communities amongst the barren stoniness of the highest country. Amongst these hardy survival types which are also of quite exceptional small beauty are the scented star-flower, creeping buttercup, alpine plantain, alpine willowherb, and the insect eater alpine sundew. The sundew's leaves are covered with fine dark hairs, each tipped with a sticky liquid. An insect alighting on the hairs becomes stuck in them and is soon completely enmeshed. The plant then produces a liquid which dissolves the insect so that it can be absorbed through the leaves.

Apart from the winter snows, alpine plants have to survive movements and slides of rock and soil, large temperature changes every twenty-four hours, and are subjected most of their lives to the snatching, beating, smashing impacts of gale-force winds, high velocity hail, and as well, the abrasion of windblown sand. Despite the apparent fragility of their large-leaved flowers, these plants are tough and tenacious of life, as are all the plants that make up the alpine communities amongst the boulder fields and gravels and screes on the mossed and lichened fastnesses of the high slopes and ridges. Baron von Mueller is commemorated in Mueller's buttercup with its brightly yellow but smallish flowers, and Australian edelweiss and snow buttercups are prolific in the herbfields of the heathy summits with their bog pools and stony crystal streams.

As the upper slopes slant down, small shrubs, often brightly flowered like the scarlet *Grevillea victoriae* appear, and with them the gnarled and twisted trunk and branches of the alpine snow gum, which reveals itself more truly further down as a quite tall, straight trunked, handsome tree growing in a group with some other snow gums. Here too, in an apotheosis of adaptation to hostile conditions is the one species of wattle growing high up, the alpine wattle, flowering in the hanging valleys of the hills.

Below the snows in all the innumerable lower-altitude environments, there are flowers, and varieties and species of flowers, impossible still, after about 200 years of settlement, to grasp and record. It has been estimated that one small sandstone area in New South Wales alone, has more than 10 000 different species of wild flower growing in it! The rainforests grow a profusion of orchids, delicate exotic flowers incongruously named, amongst moss and fern and creeper: slipper orchid, potato orchid, bearded greenhood, pencil orchid, and tree spider whose stems are rectangular in shape! Nevertheless, the majority of native wild flowers are the dry-loving xerophytes which spread and prosper on the worst of the sandy soils from which they manage to extract and make use of the mineral nutrient that is present.

From north to south down eastern Australia there are the tropical rainforests of Queensland and New South Wales, the warm temperate rainforests of the coasts of southern New South Wales and Victoria, and the cool temperate rainforests found in the west of Tasmania and in the highlands of southern Victoria. Each area possesses its unique plants. In Tasmania there is a variant of the heath family which grows up to 13 metres in height and is known as the grass tree. The Tasmanian rainforest is notorious amongst geographers and explorers as the most difficult jungle in the world because of its horizontal scrub, and *Bauera*. The horizontal scrub is made up of a plant which grows a thin stem swiftly to a considerable height. This then collapses under its own weight into the horizontal at ground level, where more stems grow upwards from the original until they too collapse and the process repeats itself until an impenetrable mat of stems, several metres in depth, is woven. A person falling down through horizontal scrub has very little chance of ever getting out again. Ironically, the plant is one of the saxifrage family, which are usually tiny decorative alpine plants. The *Bauera* is in fact like a rose bush. It grows profusely, and toughly, despite its pretty pink and white flowers, to a density almost impossible to penetrate.

Elsewhere, beyond these damp pockets in the generally dry soil garment of Australia, the xerophytes, by their hardiness, have come to dominate the remaining landscape. The omnipresent eucalypts and their cousins the tea-trees are accompanied by two other big families. The first includes the wattles. Beyond these families grow the common wild flowers such as small-leaved boronia, flannel flower, Christmas bells, bottlebrush, match-heads, bitter pea, running postman, and *Leptospermum.*

The drier areas are open forest areas, mostly of eucalypts in a variety including the ironbarks, messmate, stringybark, and mixed amongst them many varieties of acacia. Where rainfall drops sharply towards the interior of the continent the eucalypts still manage to go on surviving but alter their shapes in order to do so. Instead of branching out from a single trunk they branch from ground level. This way of growing ensures a self-made water-conserving mulch around the roots, and has other advantages. Dry country eucalypts are known commonly as mallees.

The majority of wattles are bushes. The few tree types are almost always found in eucalypt forests in damp ground. There are over seven hundred varieties in Australia, and the genus properly named as Acacia, propagates through seed pods, and has extremely hard-coated seeds which can often only be germinated through the extreme heat of fire. Most wattles begin life with small, delicate feathery leaves, but these change as growth continues into much larger, broad, flat leaves. Amongst the more common wattles are blackwood, golden wattle, hickory wattle and silver wattle.

For variety, the waratah, part of the *Protea* family, is fascinating. The waratahs themselves are distinctively limited. Victoria has only one variety, the Gippsland waratah, Tasmania has only one, and New South Wales has two varieties. Both the Victorian and Tasmanian waratahs are different in appearance from the New South Wales varieties. Other members of the *Protea* family are the banksias, the woody pear, the mountain devil or honey-flower, the spider flowers or *Grevillea,* the smokebush, and the silky oak of Queensland and New South Wales.

Ferns are amongst the most ancient of the world's plants, and Australia has varieties ranging from tree-ferns 20 metres in height to small, dainty unobtrusive plants almost invisible underfoot. The ferns don't produce seeds but instead spread spores from containers on the back of their leaves. Some spread by putting out runners. Most ferns are found in cool, moist, shady places along creeks and streams and in the rich dampness of deep rainforest gullies. A few grow in running water. Fern leaves are intricately divided in order to gain the maximum of light in their usually heavily shaded situations.

THE AUSTRALIAN BUSH is fast disappearing. Commercial exploitation, urban development, mining, land clearance, toxic chemical spraying, the spread of trail biking and the availability of four-wheel-drive vehicles, as well as thoughtlessness, carelessness and recurring big destructive fires, usually of human origin, have all put the remnants of the bush at risk. Pests too, thriving in the ecological unbalance that has been created, are playing a part in the destruction of the bush. Insect plagues, plant parasites, wood borers, have decimated large areas. The mulga scrub, once dense and almost impenetrable, is mysteriously dying out.

The richly varied and enchantingly beautiful cover that evolved over millions of years to clothe the continent has been reduced to a threadbare poverty in less than 200 years.

NATIONAL PARKS AND NATURE RESERVES are slowly spreading, but the demands made on them, particularly near large cities, are so great that new problems are already arising. One of these comes from the provision, in recent years, of nature tracks. These are already getting so much use that the tracks and their accompanying vegetation are being worn out by the numbers of people passing over them, and new tracks, involving the destruction of yet more bush, are having to be created. City populations, increasingly frustrated by the pace, pollution, and unnaturalness of their concrete and bitumen lives and driven by deep psychological needs, have unwittingly begun to devour the bush they enjoy and want to preserve. The future can only bring a development of this latest threat in a quickly increasing progression that is bound to destroy what remains of the bush.

A small part of the solution is an increased awareness of the need for conservation and rapid practical implementation by governments. The larger part of the solution is a change from the great-city patterns of living and a more natural lifestyle which would diminish substantially the demands, growing daily, that urban people are making on our most precious natural resource, the bush.

Further Reading

Charles F. Laseron, *The Face of Australia*, Angus and Robertson, 1969.

H. and A. Learmonth, *Regional Landscapes of Australia*, Angus and Robertson, 1971.

I. Holliday, *A Field Guide to Australian Trees*, Rigby, 1976.

N. McLeod, *Australian Wildflowers*, Door/McLeod, 1984.

J. Brownlie, *Australian Wildflowers*, Curry O'Neil, 1982.

D. Newman, *Australia's Native Trees*, Child and Associates, 1989.

L. F. Costermans, *Trees of Victoria*, Author Publisher, 1981.

CHAPTER 9

WALKING AND WHERE TO GO

People seem to think that there is something inherently noble and virtuous in the desire to go for a walk.
MAX BEERBOHM

WALKING is the economical, independent, healthy art of getting from place to place without the assistance of machinery. Walking is a traditional art, practised and perfected by our ancestors but now like so many traditional arts, falling into desuetude. Yet walking is the best all-round form of exercise and can be adjusted to suit most age groups and states of health. It also cures acute vehicle dependency feelings. Walking is the best way of *experiencing* the country, of re-establishing contact with the soil and the nature from which we come which is our true mental, physical and spiritual health. It is the best, simplest and most efficacious of recreations, but like most neglected arts it needs some thought and preparation before it can be properly practised.

Ancient arts demand ancient proverbs, and make haste slowly, is the one for walking. Feet and legs unused to walking more than a few metres at a time are soft-muscled. Tone has to be built up again and this means taking short walks as often as possible, in fact as often as one would unnecessarily drive the short distance to post box, or local shop, or the few-streets-away friend.

SHORT WALKS are not only important physically in preparing muscles for the pleasures of free, rhythmical walking over longer distances, they are also a useful psychological preparation: walking, after continual riding, is like entering an unfamiliar element with unexpected demands and rewards. Weather becomes a reality again; the sky something to be consulted before setting out; distance begins to turn into time taken; sight and hearing supply surprises. There is the time and opportunity to pause and rediscover the fascination of the variety and detail of a world previously only glimpsed in passing.

We all walk in our own way and have our own style, which is unique. There is no best way of walking, only one's own way which improves with practice. We are not machines, although we may often behave like them. Finding our best speed, which is the most comfortable rhythm that enables us to cover reasonable distances while still able to look and listen and *be part of* the natural world with its endless fascinations, is not like selecting one of a number of gears in a car. Individually, we have only one best pace and we have to find it and accept the limitations our bodies will inevitably impose as accompaniments. Having found our best, most comfortable pace, it is then quite easy, with practice, to quickly extend the distances covered. Yet even a best pace is not infinitely extensible. There are limits to the distance any of us can walk in a day, and if the country is at all hilly the straight line distance will be much reduced and the time increased. Working against gravity simply carrying the weight of one's own body puts a big strain on heart and lungs and muscles. Carrying a pack weighing only a few kilos on one's back greatly increases the strain and inevitably reduces the speed and distance covered.

WHEN PLANNING A WALK, look out first of all for the contour lines of hills and make a generous time allowance for them. A number of formulae have been worked out to make allowance for climbs and descents but like all formulae they are merely useful generalisations and have to be substantially modified to meet the needs of individuals and groups. Experience is the best generator of one's personal time allowance. Hard experience will quickly prevent any tendency to euphoria and over-optimism. It is always better to plan for a little less, than to plan for the little more that produces over-tiredness, disappointment and irritability. In bad weather conditions in high country, an unrealistic approach to distance planning can end in disaster and loss of life. Make allowance too for the sort of country you will be walking through. There are all sorts of paths and tracks and trails through the bush which are usually marked on good maps. If the maps are more than a year or two old great changes may have taken place: paths may have disappeared, tracks become overgrown, trails obstructed by fallen timber. A whole new set of trackways may have obliterated the old ones marked on the map. It is important to know whether obstructions are likely to slow you down, whether there are large impassable areas of bog or dense bush that will have to be outflanked, and whether the tracks you are following *are* the ones shown so clearly and neatly on the map.

WALKING TACTICS are of the most simple sort and therefore quite difficult, especially for young people. When setting out on a walk amble along *slowly*. Take your time while muscles warm up and the body adjusts itself to movement. If others want to rush on ahead, let them; the chances are that you will pass them at a comfortable, easy pace, without any feeling of strain, within the hour as they sit 'admiring the view'. At the end of the day, too, you will be very much less tired.

Walking is a matter of comfortable, individual rhythm, but each day has its own larger rhythm as well, the rhythm of stopping to rest and starting off again. There are no mechanical formulae to be applied here either. Rest stops are a matter of personal judgement. So long as the walk goes comfortably forward towards its conclusion and people are *enjoying* the walk, it doesn't matter how many stops are made or for how long.

Good use should be made of stops to stretch out, relax and rest completely. Feet need care and stops are the places to give it. Take the boots right off if there's time. Wriggle your toes in the cooling breeze, or soak them in a creek. Look out for incipient rubs or blisters that can be dealt with, before they get nasty and incapacitating, with first aid plasters. Enjoy the stop. Make sure that boots and clothing and rucksack are perfectly comfortable before setting off again. On a military exercise with a party pressed for time, or when reluctant children and adults have been pressed into action, a regular stop of say five or ten minutes in each hour may be necessary. Convicts in chain gangs and parties of slaves and slave traders may also need this sort of treatment. People walking for enjoyment don't. On the other hand people walking for enjoyment are sensible characters who know better than to spend the day rushing in all directions in short frenzied bursts of activity which are exhausting for them and wearing to everyone else.

WALKING UPHILL also needs the application of commonsense. The steeper the hill the more slowly it has to be tackled. Rushing it, even if you have the energy available, only makes for exhaustion and even possible physical damage. Patience is the thing. Rest a second or two every step, or every other step, if the way is very steep. In the open it may be easier to zig-zag slowly upwards across the steepness to ease strain. The distance will be greater but the comfort is appreciable. If your heart is jumping about in your chest like a terrified rabbit, your lungs are gasping on furnace-hot air, and your legs feel that gravity has increased five times, then you are trying to do too much. It shouldn't be necessary to state the palpably obvious, but on every walk involving steep hills there is always at least one member of the party pressing on heroically in exactly this way. Goodness knows what misplaced pride, impossible idealism, plain masochism, deep neurosis, or ineradicable vanity produce this painful travesty. It is an aberration to be seen in others but *never* to be indulged in for oneself.

LONG CLIMBS are best punctuated by frequent, adequate rests. This is the best and really the only way to deal with the decline in morale as the upward way lengthens and yet another crest rises up beyond the one just conquered. Admire the view. Think beautiful or philosophical thoughts. Speculate on the origins of man or the absurdities of civilisation, but keep slowly moving. This is the moment to reach into pocket or pack and get out the instant energy glucose tablets, or the Kendal Mint Cake, to linger over the special high altitude (anything steep from 150 metres in height will do) deliciousness that the rarified air and leaden limbs impart to them. Quickly available sugars lift morale more quickly, and get people up hills and mountains with greater despatch than any other single thing.

NIBBLING AND SIPPING are important contributions to good walking. Travelling up hill and down dale for long hours takes a great deal of energy which has to be recovered in the form of a regular, small intake of food. Nibbles are a personal matter and can consist of almost anything: chocolates, sweets, oat-and-honey biscuits, cheese and dates, nuts and raisins, dried fruit (snap-dried Queensland bananas are excellent, light in weight, nourishing and tasty, if you can get them) or Scroggin, which is a mixture com-pounded to suit personal preferences, usually including raisins, sultanas, candied peel, dates, chocolates, and a mixture of unsalted nuts. Very salty foods should not be included amongst nibbles. A raging thirst on a warm day when drink is scarce and the next waterhole is far ahead has to be endured to be believed. 'Little and often' is the best way to nibble. This ensures that energy is steadily available and that stomachs do not become overloaded and uncomfortable.

Water should always be carried and sipped sparingly throughout the walking day. There are now very few open water supplies free from the risk of serious contamination. In general it is better not to trust the water you find unless you have purification tablets in your rucksack. Water, or the fruit juice or cordial you prefer, is thus essential. In summer the risk of heat exhaustion and serious dehydration is much increased, and extra drink should be carried if it is not possible to plan the walk via a water supply that is known to be safe.

SOLITARY AND SOCIAL WALKING are very different things. For the person walking alone in the bush, dangers are greatly increased. A simple badly sprained or broken ankle carries with it the risk of death from exhaustion, exposure, starvation or all three together if there is no one in the vicinity and help is too far away to be reached. Of all people, the solitary walker (if he *must* walk alone) has the greatest obligation, to himself and to those who may have to mount a search-and-rescue operation for him, to leave with some responsible person the fullest possible details of the route he intends to take, the stops where he will tent overnight, and when he can be expected to return home. The simplest, most casual incident in the course of a solitary walk can have lethal consequences. Even on a day walk the solitary walker should leave behind notice of his intention if the country or bush he is to pass through is rugged or dense.

Two competent, properly equipped walkers walking together

reduce the risk and dangers by much more than half. In really rough, rugged, difficult country three is the only reasonably safe minimum. If accident or injury then befalls a member of the trio, one can go for help while the other looks after (feeds, keeps warm or cool, gives psychological support) the casualty. In the event of two people being involved in an accident there is still one left to go for help. The likelihood of all three being seriously involved is extremely remote, but if the likelihood seems a possibility then perhaps prayer mats should be included amongst the equipment!

Any group of more than three or four walkers is a party and the possibilities and consequences are much more complex. Social demands are greatly increased. Anyone joining a party of walkers should think clearly about the obligation undertaken. If you go with a party there are a number of clear responsibilities involved in your decision. Some of them are: if you set out with a party you go through to the end of the walk with it. People who change their minds and destinations and go wandering off on their own are a nuisance and a danger to themselves and the party. If you are a slow walker don't obstruct other people on narrow trails and don't worry about being left behind: properly organised parties move courteously at the pace of the slowest member. If you are a fast walker don't rush ahead and lose touch with the party, and don't be a nuisance trying to cajole, or provoke other people into keeping up your pace. If you have to carry a miniature transistor radio, use the earpiece to listen to it. Most people go walking to get away from things like transistor radios. Organised parties have a leader. This is a sensible arrangement and prevents extended committee meetings and time-wasting arguments whenever a decision affecting the party as a whole has to be made. Never argue with the leader. Support his decisions, whatever you may think privately. Give the leader any help you can. His or her job is not an easy one and involves a lot of responsibility and sometimes a lot of worry which has to be concealed from everyone else.

Parties are at a disadvantage compared with small groups who know one another and their capacities well. The party's options are limited by the people who make it up, the clash of temperaments, the inevitable feeling of irresponsibility, even if only slight, that overcomes people travelling in large groups with a recognised leader. There is an innate tendency for large parties to fragment, to drift, to leave the difficult jobs to others. In an emergency these are the basic tendencies that have to be overcome swiftly. Safety demands that the party stay together, that responsibility and duties be properly and fairly shared, and that the leader be solidly and purposefully supported.

If all this seems too much to invest in a walk, if vanity will not submit to leadership, or egotism to consideration for others, or raw individualism to (a very loose) collective discipline of responsibility then don't go with a party. It's not for you. However, if you do feel like this it would probably do you the world of good.

OBSTACLES can usually be got around in one way or another. Dense bush can be by-passed, fallen timber outflanked, bogs circumvented and rivers crossed. Rivers can be treacherous and require a lot of extra care if they are at all swift, or deep, or both. In summer especially, the stones of riverbeds become slimy with growth and very slippery. A fall, exaggerated by the abnormal weight of pack on back which alters balance, can lead to serious trouble if the momentarily helpless walker is swept into deeper water. The best support underfoot is the shingle or gravel which usually occurs anyway where bars of small stony material have been deposited at an angle by the current. Large stones, rock slabs, and mud bottoms are likely to be dangerous. Innocent looking sand bars extending into and along the edges of runs may be quicksands. These can engulf you at terrifying speed and are almost impossible to escape without help.

The best crossing places are usually where the river is broad, and relatively shallow, with a slow current. It is always worthwhile to survey the river for safe, easy crossing places for a considerable distance on either side before attempting to get across. Look for the quiet places. A bar will often account for the major part of the crossing. Always respect rivers. If in the least doubt use the length of rope in your rucksack to belay the first person to attempt the crossing and the others subsequently if necessary. Don't take boots or shoes off, they give grip on the bottom and protect the feet from injury.

If the water is too deep for fording, choose a place to swim where the current is slowest and will carry you with the least effort diagonally across to the other side as you travel downstream. Make sure that the far bank you have chosen to land on is low, easy to get onto, and extends for a safe distance. In the unlucky event of being swept past a short landing place you may not be able to get up the bank for a considerable distance.

In fording it is best to keep the pack on the back for the added stability its weight gives, especially in rapid current. In swimming, the pack (if buoyant) is best pushed, or held in front. If not buoyant a groundsheet, piece of heavy duty plastic or extra strong plastic garbage bag can be used to enclose the pack and make it float. Some people always carry something of this sort which is useful for other emergencies. Keep a sharp lookout always when fording or swimming for underwater snags or dangerous debris. If the river is too difficult or dangerous to cross at the point at which you have arrived, work your way along the bank upstream until the river narrows and shallows and the crossing becomes possible. Don't take any unnecessary risks with rivers.

T HE ART OF WALKING, like any other art, needs thought and practice. It has to be begun sensibly and slowly over short distances to begin with. For beginners, or for experts who have had a long lay-off, the best way to get going again is probably by going along to a few local orienteering events. These are usually graded into four courses, from the short and easy to the quite long and difficult, and take one into a variety of bush situations while giving the opportunity to brush up navigation techniques. There isn't a better or more pleasant way of getting into the swing of things again in congenial company.

In the south and east of the continent there are still great areas of bush and mountain, mostly quite close to the major cities, in which to camp and bushwalk, and find one's way back into contact with the land and natural things. In recent years there has been a slow growth of a sense of legislative responsibility by State governments towards the natural resources of their areas. This has been brought about largely by public pressure as consciousness of the great dangers of destroying our natural environment has grown.

The result has been that there now exist all over the continent governmentally declared parks, national parks, nature reserves, and sanctuaries. These are especially abundant in the south and east, including Tasmania, and are gradually being extended while new parks are being initiated.

In the whole of Australia there are only two long-distance tracks, the one well established, the other completed relatively recently. These two tracks are the Cradle Mountain-Lake St Clair track in Tasmania, and the Alpine Way in Victoria.

T HE CRADLE MOUNTAIN-LAKE ST CLAIR TRACK is in a national park of over 1300 square kilometres in area which includes some of the island's finest mountain scenery and Mount Ossa, 1600 metres, which is the highest mountain in Tasmania. The overland track is 80 kilometres in length and usually takes about five days to traverse. There are huts equipped with bunks at the end of each day's walk, but these are usually overcrowded at peak holiday periods when a tent should be carried. The huts are unmanned and unfurnished except for the bunks. The average time walking between huts is five hours. The track, which begins at Waldheim and runs southwards to end at Cynthia Bay at the southern end of Lake St Clair, is well marked and offers some of the finest mountain walking in Australia. However, mountain weather is always difficult to predict and sudden hazardous changes occur swiftly. Rain and snow are common in the summer months, with accompanying difficult conditions, and hypothermia (exposure) is always a risk. Anyone tackling the track should be in good physical condition and properly equipped. There are park rangers at Waldheim and Cynthia Bay whose expert advice on conditions along the route should never be

ignored. Walkers are expected to leave a route schedule in the park
log book before setting out, to record progress at huts along the way,
and notify their departure to the ranger before leaving.

THE ALPINE WAY is a 400 kilometre walking track north-eastward
from Walhalla in southern Victoria to Cowombat Flat on the
border of New South Wales. The trail runs along the Victorian Alps,
keeping to high country, and touches the summits of a number of
the principal mountains along the way. The trail is often rough and
is marked by distinctive diamond-shaped metal markers. The trail
seldom comes near shop or habitation. The going is sometimes
arduous. Some areas have little water available. Snow covers the
higher parts of the trail in winter and, as always in mountains,
extreme conditions of heat and cold may be encountered almost
anywhere at any time. Even in late spring after a night of frost boots
have had to be thawed out at a fire before they could be worn. The
Victorian Alps can be cold, wild and wet, at any season of the year,
and parties going along the Alpine Way, or sections of it, should be
properly equipped and physically fit. There are a number of link and
feeder tracks to the Alpine Way from within Victoria. All are in the
Victorian Alpine Park which came into existence in December 1989
and links with the Kosciusko National Park. It is now possible to
walk the Alpine Way to the New South Wales border and follow
trails from there via Mt Kosciusko to Canberra.

THE NATIONAL AND OTHER PARKS include bush and outback
areas of magnificence and great natural interest, though none on
the mainland are true wildernesses, fire and other tracks being found
in the most remote and apparently inaccessible areas. There are
hundreds of parks in Tasmania and on the mainland. The following
notes are only about the largest and most varied.

THE GRAMPIANS FOREST PARK in Victoria is about 240
kilometres west of Melbourne, at the mainland end of the Great
Divide. The Grampians are sandstone mountains running north-
south for almost 100 kilometres and rising dramatically to peaks
above 100 metres from the plain of western Victoria. The mountains
have been deeply carved and sculpted by weathering, and the
scenery is impressively magnificent. The area is famous for its wild
flowers and its rich wildlife. It is a popular centre for bushwalkers,
rock climbers, naturalists, and photographers. The Grampians are
not a national park but a concessionary area of land belonging to the
Victorian Forests Commission.

IN SOUTH AUSTRALIA, the chief parks are along, or in the vicinity
of, the northward running Flinders Ranges, arid rocky red hills that
begin south of Port Augusta and curve to the north-eastward between

Lake Torrens and Lake Frome in a series of summits around 900 metres and topping 1000 metres at Mt Aleck. There are national parks at Alligator Gorge, Mambray Creek and Mount Remarkable; and at the northern end of the Flinders Ranges there is Wilpena Pound, a gigantic hollow over 80 square kilometres walled by sheer cliffs. About 4000 hectares make up the State National Pleasure Resort.

MOUNT BUFFALO NATIONAL PARK is eastern Victoria's only major park in the alpine region. Mount Buffalo itself, 1500 metres, is really an outlier of the main range of mountains. The park covers about 6700 hectares, all of the rocky plateau, and looks out onto magnificent scenery over very steep to vertical rocky walls. Camping is restricted to a beautiful area of bush around Lake Catani. There are well-marked walks, which are also unobtrusively signposted, of from thirty minutes to seven or eight hours. Mount Buffalo, which is 320 kilometres from Melbourne, is a granite mass roughly 11 kilometres long and 6 kilometres wide. Streams flow across its surface and erosion has left impressive rock monoliths, outcrops, and cap-stoned monoliths amongst the small swards of sub-alpine herbage and stunted sub-alpine bush. Ski-touring is becoming popular in winter and there are two downhill skiing areas with ski-lifts. The longest climb on the Australian mainland, 440 metres and two days to climb, is over the buttress on which the Queen Victoria Pinnacle stands. Climbing is of a higher standard of difficulty than anywhere else in Victoria because of the scarcity of holds which makes apparently easy routes into climbs of formidable technical difficulty.

The area south and east of Mount Buffalo includes Mount Bogong, Mount Fainter, Mount Feathertop and Mount Hotham and the area through which the Alpine Way passes. There is a growing public agitation in Victoria for this area, and the area on either side of the Alpine Way, to be made into a great Alpine National Park.

KOSCIUSKO NATIONAL PARK in New South Wales lies north and east of Buffalo and Mount Bogong and covers over 500 000 hectares of the Snowy Mountains. The snow areas in these Australian alps are bigger than all of the snow areas in Switzerland together, and the park includes a number of mountains, including Kosciusko which rises to over 2100 metres. There are large modern winter sports resorts at Thredbo and Perisher and the park and the areas surrounding them are famous for their trout streams and rivers. This is the only region of mainland Australia to have experienced glaciation. Over most of the park, in a vertical range almost from sea level to 2100 metres, the scenery is outstandingly beautiful and the walking is magnificent. The Snowy Mountains are a high plateau rising steeply in sheer cliffs and gorges from the west and sloping in gentle rolling uplands to the east. The plateau is seamed with many deep valleys. Weather conditions can change rapidly in the mountains and the greatest care should be taken as hypothermia (exposure) is an ever-present threat all the year round. There are huts throughout the park but since at peak holiday times these may be crowded, a tent should be carried. Walkers staying overnight in the park are expected to give details of their trip and stop-over plans to park rangers who are stationed at Khancoban, Thredbo, Smiggin Holes, Blowering, and Yarrangobilly.

WARRUMBUNGLE NATIONAL PARK is almost 500 kilometres north-west of Sydney. An area of extinct volcanoes and volcanic activity, it has rising from the western plains in strange, almost lunar beauty, mountains of extraordinary and macabre shape. There are trails cut through the park to most of the larger hills and these may be used as access routes to the wilder areas. A 10-kilometre track follows the main ridgeway through the tangled hills. Rock climbing is practised in the park as well as bushwalking and there are a number of camping areas.

THE NEW ENGLAND NATIONAL PARK in northern New South Wales is one of the most neglected and least known of the national parks. Mostly virgin rainforest, its highest summit is Point Lookout at 1600 metres. Heavy snow is not uncommon in the higher areas in winter. There are snowgums and snowgrass and other sub-alpine vegetation on the highlands. Rainfall averages 2540 millimetres per annum. The park area has only limited trails and is almost pure wilderness. Conditions can be very rough and going extremely slow especially in summer when most rain occurs. Despite the cold at high altitudes, May to November are recommended as the best walking months. The park rangers should be consulted before setting off on a trip.

LAMINGTON NATIONAL PARK in Queensland covers almost 20 000 hectares mainly of rainforest, with rich and abundant wildlife and spectacular flowering trees, orchids, and tree-ferns. The park has over 140 kilometres of walking tracks linking together the hundreds of waterfalls and viewpoints. The highest area of the park rises to almost 1200 metres with views for great distances across the ranges into north-eastern New South Wales. Lamington is especially famous for its forests of Antarctic Beech trees, many of which are extremely old. The Antarctic Beech is thought to be possible evidence of the northern extent of what was once a huge Antarctic continent.

These are only a few of the largest and most varied of the national parks in the south and east of Australia. There are hundreds more covering every possible sort of country from the mallee desert of Victoria's Wyperfeld to the tropical islands which are part of Queensland's Barrier Reef Island Parks. In addition there are the sanctuaries, nature reserves and bushland under the control of the various State forestry, electricity and water authorities, which are open to the public.

Further Reading

Higgin, Cronin and McDonald, *Presenting Australia's National Parks,* Child and Henry, 1986.
John Siseman, *The Alpine Walking Track,* Pindari Publications, 1988.
C. Warner, *Bushwalking in Kosciusko National Park,* 1988.
Chapman and Chapman, *Bushwalking in Australia,* Lonely Planet Books, 1988.
F. W. Hall, *Bushwalking in the Victorian Ranges,* Rigby, 1976.
Victoria's National Alpine Park — brochures and maps, Department of Conservation Forests and Lands, Victoria.
M. Morcombe, *Discover Australian National Parks,* Lansdowne, 1983.
Durant and Parv, *Wild Places of Australia,* Bay Books, 1983.
R. Raymond, *Discover Australia's National Parks,* MacMillan, 1985.
A. Fairley, *The Observer's Book of National Parks of Australia.*
Bushwalking in the Victorian Alps, Melbourne University Mountaineering Club, 1974.
Dick Johnson, *The Alps at the Crossroads,* Victorian National Parks Association.
Publications by the Forestry and National Parks Departments of the Australian States.

WHERE ARE YOU?

*Nature is a labyrinth in which the very haste
you move with will make you lose your way.*
FRANCIS BROWN

A compass rose showing some of the traditional names of the points but also divided up into 360° degrees.

With the exception of the harsh, dry, inland desert areas, which are not the concern of this book, the bush is omnipresent in all wilderness areas of Australia. Even in the Snowy Mountains, which rise at Mount Kosciusko to over 2650 metres, low shrubby growth is found in sheltered places. It is easier to get lost in Australia than in most other countries. Where the bush comes down to the edges of some of our city suburbs, people of all ages, from infants to octogenarians, become lost. Some have been lost for days before being rescued. Others have perished, almost within sight of their own homes, from hunger, thirst and exposure. What Australia lacks in large, fierce, predatory carnivores it more than makes up for in the extent, density, and relative featurelessness of its bush.

To move only a short distance from a main road into the bush can be a dangerous enterprise, unless you are properly equipped with map and compass and the skill to use them. To do so is foolhardy as well as inconsiderate of the many people who might have to become involved in a search-and-rescue attempt. The first demand of the bush is for self-reliance, which here means taking full responsibility for yourself and your safety by knowing always exactly where you are. If you can't answer the question 'where am I', you shouldn't be where you are.

T HE TWO ESSENTIAL PIECES OF EQUIPMENT for the bushwalker or orienteer are a map and a compass. The compass should be of the Silva type, which is both compass and protractor in one. It is made of transparent, shatter-proof plastic so that it may be laid flat on the map without obscuring much detail. These compasses are available in a range from the simplest form to the most complex, with exchangeable map scales, step-counter tachometer, large baseplate and magnifying lens, a mechanism to make allowance for magnetic variation, and the usual precisely balanced, oil-damped compass needle set in a rotatable housing marked off from 0° to 360° with orienting lines set in the bottom. A special model which is generally available, has been made for the British Army, which uses orienteering in much of its training.

Before the Silva-type compass was developed, accurate map-and-compass work involved carrying a protractor, which was often difficult to use and took an excessive amount of time. The Silva-type compass makes a protractor unnecessary and simplifies the taking and transferring of bearings from map to terrain, and from terrain to map, so that the operation is almost instantaneous. The more expensive prismatic compass used for precision survey work and mapping is not wholly suitable for orienteering, and is unnecessary for walking if you possess, as you should, a Silva-type compass of good quality.

A Silva-type compass

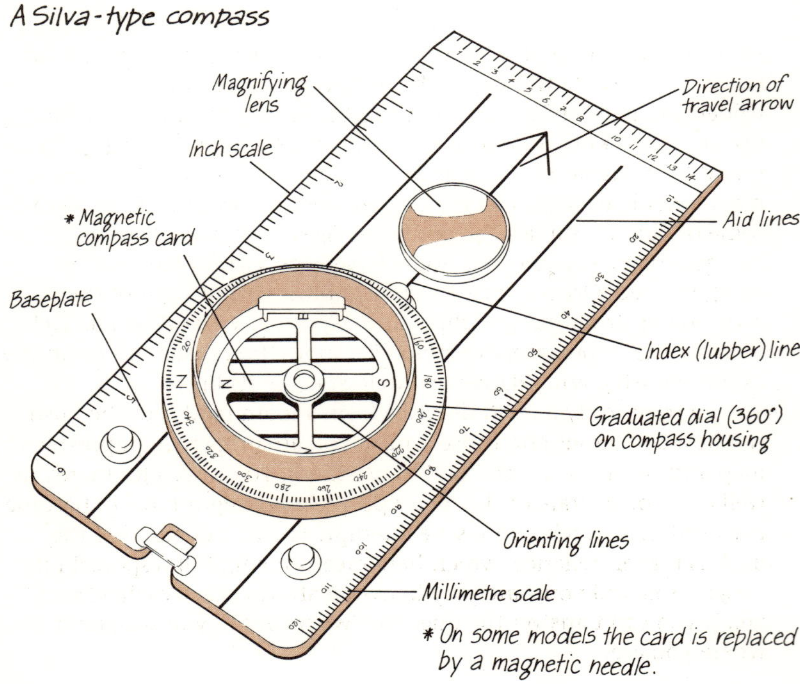

* On some models the card is replaced by a magnetic needle.

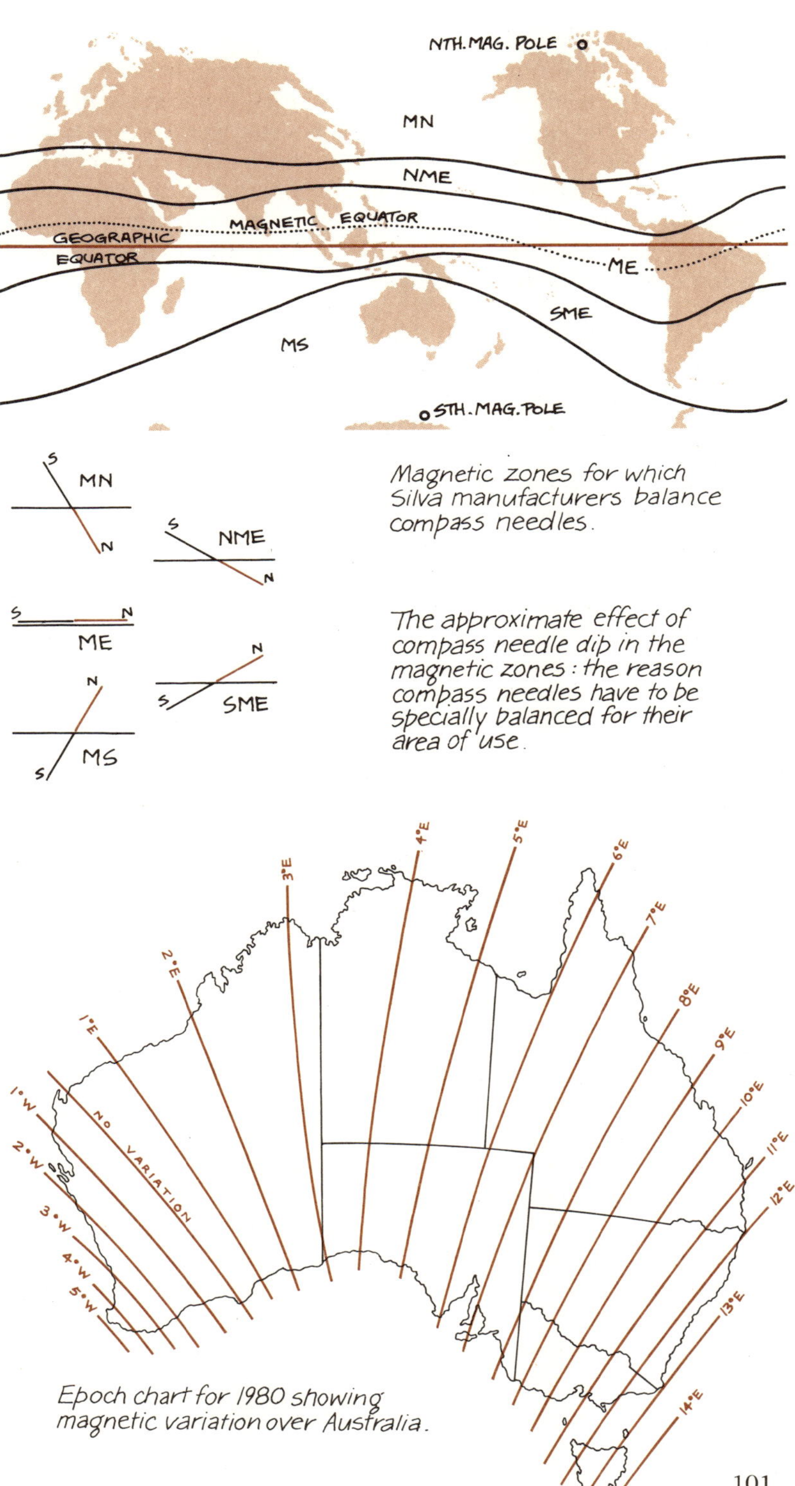

Magnetic zones for which Silva manufacturers balance compass needles.

The approximate effect of compass needle dip in the magnetic zones: the reason compass needles have to be specially balanced for their area of use.

Epoch chart for 1980 showing magnetic variation over Australia.

Because of the variation of the lines of magnetic force over the surface of the globe and the force extended by them, the needle of a compass is often pulled down at one end unless counterbalanced in manufacture. The Silva compass makers now recognise four magnetic zones and make their compasses to compensate for the dip in each. So a compass which is accurate in Britain (zone MN) will not be of use in Australia (zone MS). A glance at the map of the magnetic zones will show that compasses for use in Australia should only be used in Australia and the Antarctic!

MOST PEOPLE ARE FAMILIAR WITH MAPS from their early schooldays, familiar but not intimate; and intimate knowledge of maps is what is needed if they are to be used creatively in crossing unfamiliar ground or exploring new terrain. Fortunately maps are simple to read and understand and intimacy with their features grows quickly with practice. There aren't many things to remember.

A map is a formal picture of the ground on which signs have replaced all the familiar features.

Nearly all maps carry in their margin or at the bottom a list of signs and an explanation in words of what they mean. This list is called the *Legend*.

Whether the land rises, falls, or remains flat is shown on the map by contour lines which join together land of the same height. Some Australian maps show heights by colouring from very light brown to dark brown; dark brown showing the highest land. Contour lined maps are much to be preferred, because they show clearly the steepness or flatness of the land they cover by their closeness together or width apart. Lines close together show very steep slopes. Lines well apart show much flattened slopes. On the edge of the map the *contour interval* is usually printed. It is necessary to know this if the steepness, or otherwise, of the land is to be accurately estimated. The contour interval is the number of metres (or feet) between two contour lines. This distance is the same between all the lines on the map so that if the contour interval is 10 metres and there are five contour lines running together and almost touching there is at that point a very steep rise of 40 metres. In the same way two contour lines 500 metres apart show that the ground only rises 10 metres over that whole distance. The contour interval is the vertical scale of the map, while the scale (also printed on the edge of the map) is the horizontal scale.

Most maps use several colours in their make-up; for example, water (streams, lakes, dams, rivers, the sea) is usually shown in blue. Coloured maps are usually better and easier to work with than black-and-white maps.

Scale is the proportion between actual distances over the terrain and the distances shown by the map. A scale of 1 : 10 000 means that 1 centimetre on the map represents 10 000 centimetres or 100 metres on the ground. This is a 'close-up' picture which will show a small area of country in clear, fine detail. On a map with a scale of 1 : 250 000 one centimetre will take in much more ground (1 centimetre on this map represents 2500 metres or 2.5 kilometres) and there will be less room left to put in fine detail. This could be compared to a picture made from much higher up, from a high flying aircraft perhaps. Higher still, the picture from an orbiting satellite covers much more ground again. Actual satellite photographs show the outlines of countries long familiar from atlases, and are, incidentally, a strikingly beautiful confirmation of the precise work of generations of map makers and geographers. The scale at this height is now about 1 : 1 000 000 or 1 centimetre on the map equal to 10 kilometres. A typical world map, spread over two pages of a large atlas, would have a scale of about 1 : 48 000 000 or 1 centimetre equal to 480 kilometres, when only very large details like seas and great mountain chains can be recorded.

The two scales most common on maps made in Australia by the Department of Minerals and Energy are 1 : 250 000, and 1 : 100 000. On these maps 1 centimetre can be taken as 2500 metres (2.5 kilometres) and 1000 metres (1 kilometre) respectively. Neither is really suitable or useful for the walker because they cover much too large an area in too little detail. Maps useful for walkers begin at a scale of 1 : 50 000 (1 centimetre equals 500 metres).

Orienteers, who are interested in particularly precise navigation over relatively short distances, prefer maps with scales of 1 : 25 000,

Landscape in perspective and in contour lines.

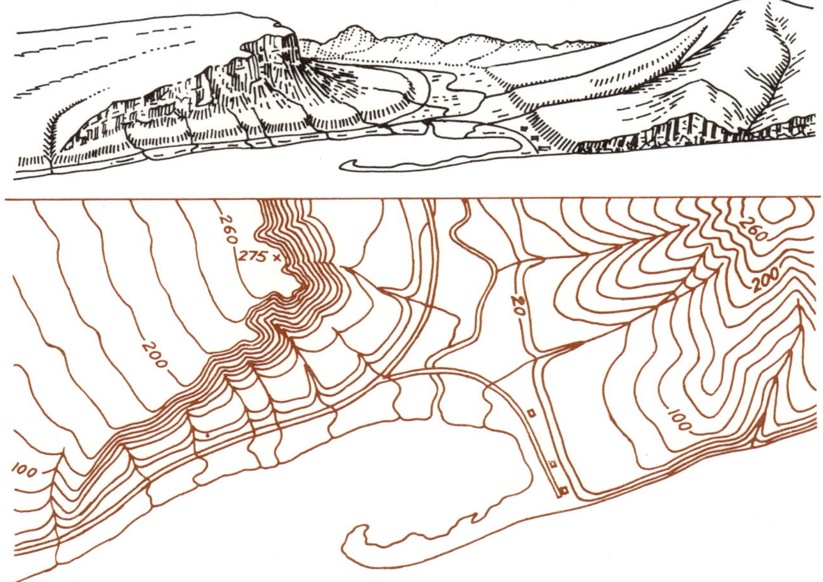

1 : 20 000, 1 : 15 000, or even
1 : 10 000. On these maps the
detail is so fine that individual
trees, ant hills and large stones
can be recorded.

Compasses for orienteering,
bushwalking and outdoor work
generally, are useful because
they can show how we stand in
relation to the land around us,
and enable us to travel in a
relatively straight line to our
destination.

A person standing in the
open and facing the magnetic
north indicated by his compass
needle, is always in the centre
of an imaginary circle of 360°,
with 360°/0° straight ahead. To
the right will be 90° or east,
behind 180° or south, to the
left 270° or west. The old names
of the points of the compass are
tedious to remember and use,
and not as accurate as degrees.

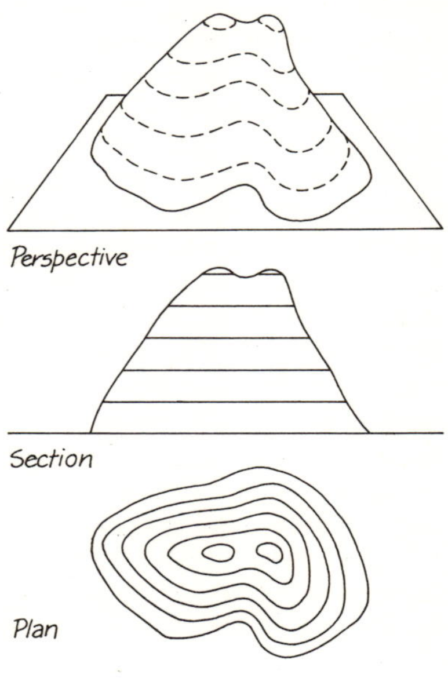

Contours give a clear
picture of the rise and
fall of the ground if
they are read properly.

As an example: behind, and to the right, almost a kilometre
away from the person in the circle, is a small steep hill. Using the old
names of the points of the compass it would be described as south-
south-east. This is only a rough indication. The direction of the hill
from where the person is standing is actually 168°, which is quick to
note, far more accurate, and needs no carrying around in the head of
a clutter of compass point names. The direction of the hill (168°) is
called *the bearing*.

TO TAKE A BEARING, or find the direction, of a distant feature is
very easy with the Silva-type compass. The direction-of-travel
arrow in the front of the baseplate is pointed at the feature. The
compass housing is then rotated until 0°, or north, is opposite the
north-pointing end of the needle. This brings the orienting arrow
outline on the transparent bottom of the housing squarely beneath
the compass needle. The bearing of the feature is then read off from
the *back* of the direction-of-travel arrow.

Unless a map can be related to the ground it covers it is of little
or no practical use at all. In clear weather the map can be set to relate
to the ground by lining up prominent features on it with prominent
actual features of the countryside that are visible. The map is lined
up with *several* features at the same time to make sure that it is

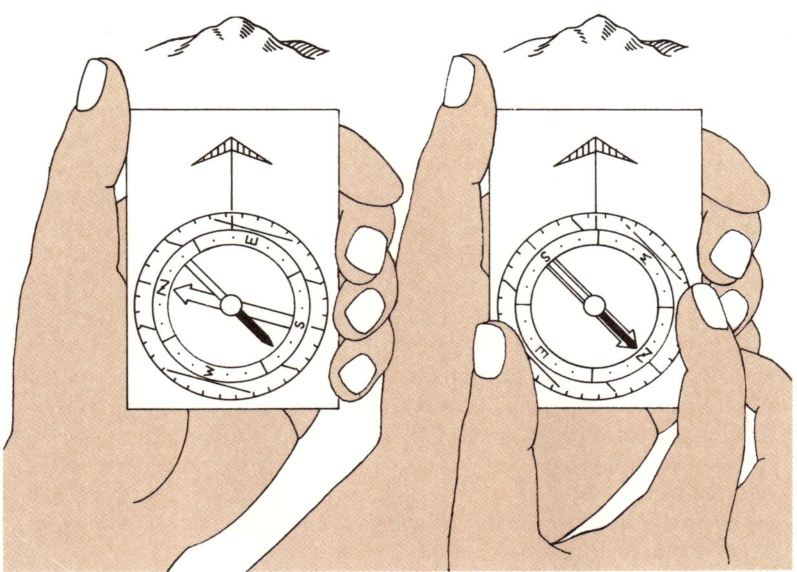

Taking a Bearing

1. Point direction-of-travel arrow at feature (hill).

2. Rotate compass housing until orienting arrow and needle coincide. Read off the bearing at inner end of direction of travel arrow. (index line).

correctly set. Using only one feature can easily lead to dangerous error if the feature is mistaken for a similar one nearby. The use of several features enables cross-checking to be carried out and the risk of error is greatly reduced.

A quicker and better way of setting the map, and the only way in bad weather, or in poorly mapped or featureless country, is to use the compass. The magnetised end of the compass needle (usually coloured either black or red, and sometimes marked with luminous paint) always points to the magnetic north pole and will go on doing so while the compass is held level, unless ferrous metal (wristwatch, fence wire, pocket knife, sheath knife, belt buckle, some rucksack frames: the list is endless) is brought near to it.

Recent Australian maps have printed in the margin three arrows which indicate north; that is, to the actual north pole, to magnetic north, and to grid north. In this case it is a simple matter to put the compass on the map and turn it until the north-pointing end of the compass (make sure it *is* the north-pointing end and not the south-pointing end) and the magnetic north arrow on the map indicate precisely the same direction when the map will be set towards magnetic north. Before finally adjusting the map to true north some other factors have to be considered.

The position of the north magnetic pole and consequently the magnetic lines of force around the planet are changing all the time, but at known rates. The magnetic north arrow on a map is usually only *precisely* accurate for the year in which the map was issued. After that it is necessary to add or subtract degrees, or parts of degrees, to make it entirely accurate. Often this annual variation is so small that it can be ignored altogether. As a general rule the older the map (all good maps carry their dates of first publication and subsequent revisions) the larger the correction that will have to be applied.

In Australia we are fortunate in that our map makers use the grid system of map reference which is also current in Britain. With this system it is possible to refer exactly to any spot on the map by using a six-figure map reference. On all our maps there is a network of fine black lines forming a grid of squares over the map area. The lines forming the grid are numbered vertically and horizontally at the edges of the map. All that is necessary to locate the feature to be referred to is

(1) Note the number of the vertical grid line to the *left* (i.e. the west edge) of the feature.

(2) Estimate in tenths of a grid square edge the distance from the vertical grid line to the feature and add it to the grid line number. This will give a three-figure number.

(3) Note the number of the first horizontal grid line *below* (i.e. the south edge) the feature.

(4) Estimate in tenths the distance from the horizontal grid line to the feature. This will give another three-figure number. The grid reference is the combined six figures (vertical reading followed by the horizontal reading).

All the vertical grid lines run from south, to north at the top of the map. For various reasons the north of the grid lines is slightly different from true north, but for all practical cases it can be considered the same. However, the arrows showing true, magnetic and grid north in the margin of the map should always be examined

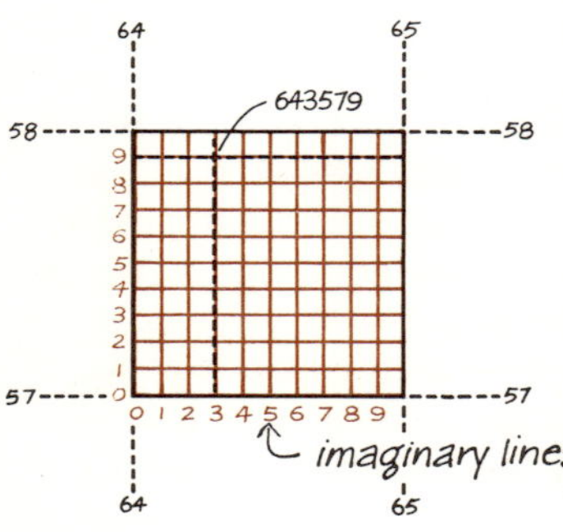

East : Take the west edge of the kilometre square in which the point lies and read off the large figures printed at the ends of this line in north or south margins :   64
Then estimate tenths eastwards :   3
                                    643

North : Take the south edge of the kilometre square in which the point lies and read off the large figures printed at the ends of this line on east or west margins :   57
Then estimate tenths northwards :   9
                                     579

to make sure that the difference between true and grid norths is not excessive. It should not be much more than a degree and a half at most. Most maps have north at the top edge but a few do not. Always check if you have to work with a new or unfamiliar map.

In setting the map we have seen that magnetic north is varying from true north all the time, and that grid north varies fractionally from true north, but can be considered as true north in practice. The final step in setting the map (lined up with magnetic north) has now to be taken. From here on only two norths will be considered, grid (approximate true) and magnetic.

At present the grid lines are pointing to magnetic north and have to be aligned with true (in this case grid) north before

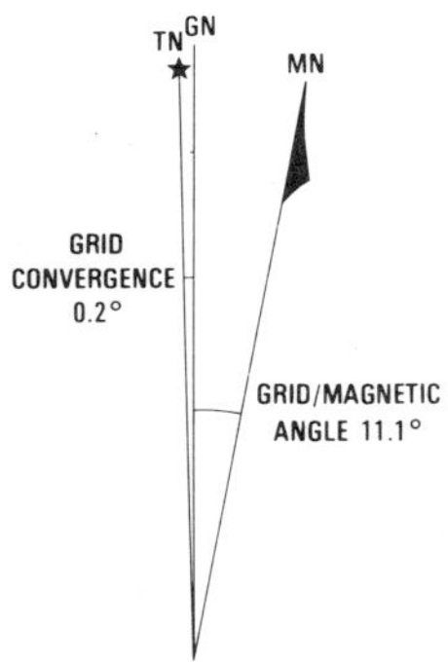

True north, grid north, and magnetic north: magnetic north is correct for 1980 and moves easterly by 0.1° in about 3 years.

the map finally corresponds with the surrounding country. A look at the margin shows that the difference between grid and magnetic north is 11°E (i.e. magnetic north is 11° to the east of grid north). Here the map has to be moved around 11° to the west before it is correctly set. This is very confusing but not something to worry about. Along the whole of the eastern side of Australia magnetic variation is eastward so an easy mnemonic is possible. This is the word MAGS. MAGS stands for two pieces of information: if working from *Magnetic Add,* if working from *Grid Subtract.* (And the opposite is true in areas with a westerly variation. See Epoch Chart.)

In this example we have been working from Magnetic so we add 11°, to the 0° of the magnetic north of our compass and adjust the map (to the west) accordingly. All of this may sound very complicated on paper but with a Silva-type compass, both this and the reverse operation, when working from map to country, are simple, quick, and easy to do.

The best of all solutions to this problem of the difference between magnetic and grid north is a sublimely simple one: prepare your map(s) by drawing in *magnetic* north lines before you set off. From then on ignore everything except your compass and your magnetic north lines.

It is very easy, using an ordinary school protractor, or better still a protractor compass, to cut out a cardboard template of the number of degrees of difference between grid and magnetic north for each of your maps, most of which will differ from one another in their variation. Lines can be drawn in at regular intervals across the map, preferably in a distinctive colour. Dark green seems to work well on most maps. Magnetic variation is usually only a small fraction of a degree annually so that your prepared map will probably be good for

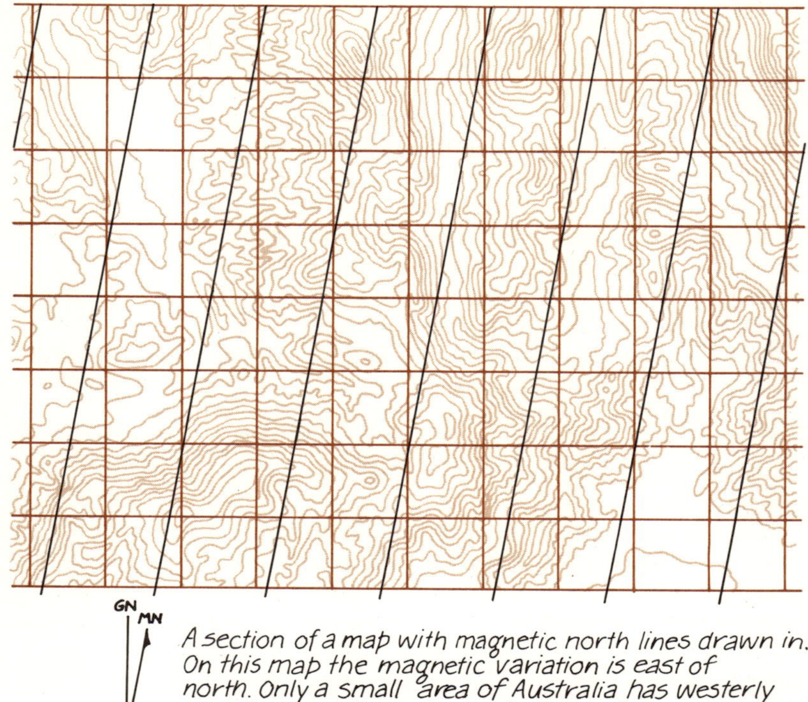

A section of a map with magnetic north lines drawn in. On this map the magnetic variation is east of north. Only a small area of Australia has westerly variation and this is in the far west of the continent.

ten years or more. Working solely with magnetic bearings in this way saves time, effort and confusion. Worrying about whether you should add or subtract *x* number of degrees in the middle of torrents of rain, in over-century heat, at the end of a long and exhausting day is quite unnecessary. Worry also introduces a serious risk of error. Unless you are accustomed to, and good at working with figures, even under conditions of extreme stress, it is much better and safer to spend a few minutes in the relative peace and quiet of home preparing your maps.

Even simpler and easier is the screw adjustment on the more refined Silva compass which can be used to make allowance for the magnetic variation. Peace of mind, which leaves freedom to enjoy being outdoors isn't always as easy or as cheap to purchase.

DIRECTION OF TRAVEL-MAP TO COMPASS BEARINGS. This is a practical situation which you will encounter almost every time you go into the countryside with map and compass. You are at point A and you want to get to point B which is hidden from you beyond some moderately thick bush (or across a mountain spur, or

whatever). Find point B and then point A on the map, then
(1) Lay the edge of the compass baseplate on the map so that it passes through both points. Make sure that the direction of travel arrow points *from* A and *towards* B.
(2) Turn the compass housing so that the orienting arrow and lines coincide with the magnetic north lines you have drawn on your map. Make sure that the orienting arrow and lines are pointing to magnetic north. (A common mistake is to have the orienting lines coinciding with the map magnetic north lines but the orienting

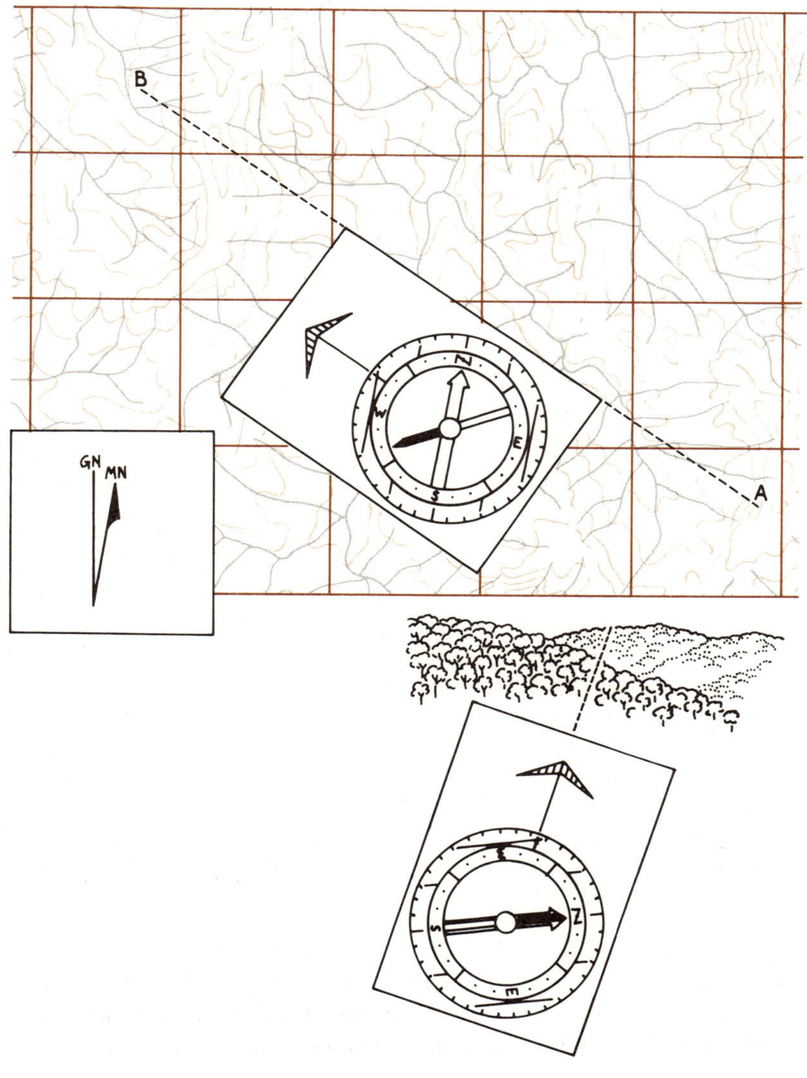

arrow pointing *south*! This could send you, albeit somewhat puzzledly, in the opposite direction to that in which you wish to go.) (3) Take the compass in your hand with the direction of travel arrow pointing away from you and turn *yourself* around until the needle points to north. The direction of travel arrow will now point in the direction of point B. Also at the inner end of the direction of travel arrow you may read off the bearing of point B if you require it. If, having got to point B, you want to return directly to point A there is no need to add 180° to the bearing on your compass and so reverse your direction. All that is needed is for the compass to be turned around so that the direction of travel arrow points *towards* you and then follow the direction indicated by the opposite end of the baseplate. This is much easier than adding 180° to your original bearing with the risk of miscalculation and subsequent error.

If the map has not been prepared with magnetic north lines beforehand the procedure is slightly more complicated. Move (1) remains the same. Move (2) remains the same except that the orienting arrow and lines are set to coincide with the *grid north* lines and therefore allowance must be made for magnetic variation by adding or subtracting the necessary number of degrees. Here we are working from Grid and so Subtract by turning the compass dial the required number of degrees. Move (3) is carried out exactly as before.

COMPASS TO MAP BEARINGS. It is often necessary to take a bearing of some distant feature and transfer the bearing, as a line, to the map. To do this face squarely towards the feature, holding the compass so that the direction of travel arrow points towards the feature. Turn the compass housing until the magnetic needle points to north (bearing 360° on the dial). Lay the compass on the map so that one of the longer edges of the baseplate cuts the object or feature and the orienting lines of the compass coincide with the magnetic north lines you have drawn on the map, making sure that the orienting arrow is pointing north and not south. A line drawn from the feature along the edge of the baseplate is your bearing. Incidentally, your own position is somewhere along this line, a useful piece of information as will be made clear later.

If your map has not been prepared and has only grid north lines on it, adjustment must be made to allow for the difference between grid and magnetic north. Here you are working from magnetic to grid again so the mnemonic MAGS says Magnetic Add, and so you add the appropriate number of degrees to your compass dial before drawing in your bearing.

LOCATION BY BEARINGS. It often happens that you are in the vicinity of a conspicuous continuous feature (track, power line, canal, creek . . .) but unsure of your exact position along its length. Here you can take only a single bearing, as described above, and

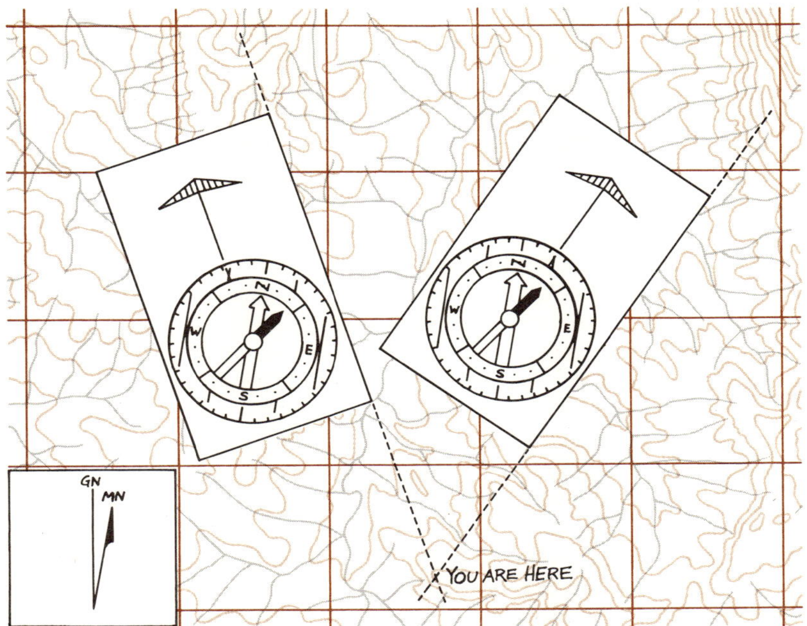

Back bearings can be used to locate oneself fairly precisely. Two distant landmarks are identified and a bearing taken from each of them. Add the magnetic corrections, plot these lines on your map and where the lines intersect is your position.

transfer the line of the bearing to the map. Where the bearing line intersects the feature is your position. If you are using an unprepared map it will always be necessary to convert your magnetic bearing to a grid bearing by *adding* (Magnetic Add) the number of degrees difference.

When there are no easily identifiable features nearby from which you can locate yourself, back bearings can be taken on two distant features, preferably separated by 90° or more of arc. The back bearings are transferred to the map and where they intersect is your position. The illustration makes this clear.

If great accuracy is required, back bearings can be taken on *three* distant features (if you can find three; recognisable features are always difficult to find when you most want them) and projected onto the map. This projection results in a triangle. Your position is within the triangle. The larger the triangle resulting, the greater the error involved in the bearings taken, and conversely.

ERROR IN ROUTE FINDING is chiefly human error. Reputable maps (you shouldn't use any other sort) are usually quite accurate even when old. Compass needles seldom point to anywhere but magnetic north, but it's a good idea to check yours occasionally. No compass will point truly to magnetic north if it is used in the immediate vicinity of a motor car, a metal building, a load of steel girders, an electricity pylon or a metal suspension bridge. Equally, it will be in error if held near to a photographic exposure meter or camera, a set of keys, a steel bracelet, or a tin mug.

In moments of stress it is all too easy to identify wrongly what seems to be an obvious feature on the ground where the map says that there ought to be one. But is the map set properly? Does the ground correspond *in detail* with the signs on the map? Can a check be made by reference to a second feature?

Error creeps in as well in estimating how far has been travelled. The map should be consulted and related to the ground at frequent regular intervals so that at any given moment you know quite accurately where you are. As a backup system to your route finding and map work you should know your average times walking over various sorts of terrain. How long do you take to walk a kilometre on a road, over grassland, through thinly bushed country, through densely bushed country? How does your average speed alter when travelling uphill in all these circumstances? With a little practice most people can quickly learn to estimate their pace over a variety of ground sufficiently accurately for their needs. Knowing how to compute distance travelled is a small but important skill. The pleasure to be had from using that skill is out of all proportion to the effort involved.

A sophisticated adjunct for the bushwalker is the pedometer, which records distance walked. A good pedometer is relatively expensive but is beyond price in its value for accurate navigation and the maintenance of safety. Knowing fairly precisely how far you have walked, together with the direction taken, and your starting point, means that you can pinpoint your position with great accuracy at any moment and avoid the errors that inevitably arise from guessing distances. Pedometers commonly work on the pendulum principle, each step taken causing a recording swing. They can be adjusted, approximately, to the length of the individual step and it is important, if really accurate results are to be obtained, to do this carefully.

Old craftsmen-boatbuilders had a maxim that ran: measure twice and cut once. It's a good maxim to apply to map and compass work. Do each operation and calculation twice and you'll only have to walk your track once. You'll also arrive safely at the destination you aimed at.

If error occurs *you* have probably made it. Look first to yourself before starting to blame the compass, the map, the weather, your companion(s), the last stray dog you saw, or the dead wallaby in the creek. If you've made a mistake, or you're in doubt about your whereabouts, don't be proud and inflexible and compound your error; *go quietly back to your last positively known spot and set off again from there.* It's much quicker in the long walk.

E VERYONE GETS LOST at some time or another. No one (well, hardly anyone) will own up to it. Getting lost is a valuable experience which should happen to everybody. When it happens, relax. Sit down, eat or drink something, stretch out and examine the possibilities. There are only three possibilities: you can stay lost, you can let somebody find you in the indefinite future, you can find your own way back to not being lost. It is a peculiarity of human nature that well over 90 per cent of human beings prefer the last solution. The other 10 per cent are people with suicidal intentions, drunks, romantics playing hard-to-get, a few simple-minded folk who think that the bush is something like the local park but without the bandstand, and an even fewer map-and-compass illiterates who haven't bothered to get themselves topographically educated.

The next question is: are you seriously lost? If you still have your map and compass you *cannot* be seriously lost. Set the map with your compass. Decide the area you are in. (You must have *some* idea how far you have travelled since you last knew where you were.) Which is the easiest and quickest way out, or to a known feature? How long do you estimate it will take?

Knowing you have a clear escape route to safety and the great unlost world if required should put you in a proper frame of mind to re-examine the situation completely and see if you can't work out a way of resuming your original route. (Just as an example: there should still be a bearing on your compass; would reversing the compass take you back to where you started from?) Perhaps your original objective is closer and more accessible than you think. If you can reach a conspicuous feature somewhere in your area is it possible to go on from there? But don't be too proud. Play it safe. If the escape route is the only logical and sensible thing to do: do it!

If you are seriously lost, that is, if you haven't map or compass, or pocket survival kit, you are in an emergency situation. Emergencies of this sort are dealt with in Chapter 12. If you have read this chapter conscientiously and carried out fully the practical suggestions in it you shouldn't be in a situation where you have to refer to Chapter 12. But you should know it thoroughly anyway, because accidents do happen, maps and compasses do get lost through no fault of their owners, and there are always a few people who were born unlucky.

Further Reading

C. Rustrum, *The Wilderness Route Finder*, Collier Books, 1973.
K. Haasma, *Orienteering*, Methuen, 1984.
J. Disley, *Tackle Orienteering*, Stanley Paul, 1984.

CHAPTER 11

BUSH BEASTS

The human being says that the beast in him has been aroused, when what he actually means is that the human being in him has been aroused.

JAMES THURBER

The Australian bush is often unexpected, but seldom frightening. Darkness is a great amplifier of sound and enlarger of the imagination. Camping in the Tasmanian bush a few years ago, we were woken in the still dark early hours by a ripping, tearing, squeaking, squealling noise, not far from the tent. Large animals scuffled together outside grunting and occasionally screaming in rage. The night filled up swiftly with the deafening demonic conflict of giant prehistoric animals newly risen from the depths of some undiscovered Tasmanian lake or tarn. Huge red eyes glowed balefully and threateningly towards the tent opening, and the ripping, tearing, squawking, squealing noise went on getting louder.

The torch beam lit up three Tasmanian brushtail possums, one biting busily on the small foam plastic box thoughtlessly left outside, in which we had left a few food remnants, the other two sitting bolt upright and staring resentfully, with their small bloodshot drunkards' eyes, at the impertinence of electric lighting. The Tasmanian brushtail possum is large as possums go, robust, thickly furred, and unbelievably stubborn. His bites still scallop the edges of the lid of our plastic cooler.

T HE BEASTS OF AUSTRALIA are different from those of most of the rest of the world. Some 60 million years' separation from other large land masses has made this inevitable. Isolation has ensured the continuous evolution of groups of beasts who died out in competition with others in the rest of the world long ago. The marsupials, animals pouched to carry their young in the early stages of life, have survived chiefly in Australia, together with egg-laying mammals like the platypus and the echidna or spiny ant-eater, which are known to biologists as monotremes. Possums are marsupials and a female with a young one in her pouch came regularly to visit the Tasmanian camp each evening while we remained there.

Australia possesses a great variety of beasts, most of which are natives largely unknown elsewhere, and which are spread out through the multiplicity of local environments of a continent stretching, in the east, from ten degrees south to forty-three degrees south. The eucalypt forests, both wet and dry, of the temperate and subtropical zones are the dominating feature of the brontosaural backbone of the great dividing range. Then there are the life-teeming tropical rainforests of the north, the widespread open woodlands beyond the forests, the flower-bright heathlands of coast and rocky range, and the small attractive herbage of the high mountains. The environments interpenetrate and mingle subtly together in mixtures that can change character unobtrusively in a few hundred metres sometimes under the influence of soil, or stream, or altitude.

The keynote of the sclerophyll bush is unobtrusiveness. To city dwellers whose acuteness of observation has atrophied, this bush is almost non-existent. Walking through it they are aware only of a tangle of closely growing trees and plants and the omnipresent fear of snakes which they imagine in terrifying multiplicity. Fear, largely groundless, and blindness, acquired from the monotony of bitumen, bricks and concrete, can make the abundant life of the bush invisible to them. The smell, the incomparable combination of sweet, sharp bush odours, poignantly, unforgettably most vivid in the cool faintly damp crispness of the dawn of a fine spring or autumn day, is hardly noticed or appreciated. Few city dwellers have been in the bush at dusk or dawn or overnight. Their impression is of the summer midday bush — hot and still and somehow threatening in its huge shadowed interior. Yet to anyone normally observant and uninhibited by fear and ignorance, the bush is always unbelievably beautiful with life, even on a summer midday of overpowering heat.

THE BIGGEST BUSH ANIMALS ARE ALL OF THE KANGAROO FAMILY and can grow up to 3 metres in length from nose to tail tip. All kangaroos, and wallabies, which are of the same family, can move at high speeds of up to 50 kilometres an hour and can jump obstacles up to 2 metres or more in height. The most common of the large kangaroos are the great greys and the red kangaroos, whose females are not red at all but a beautiful, delicate smoky blue. Wallabies, which are the smaller members of the kangaroo family, and kangaroos, are docile, quiet creatures unless cornered, and should not be approached too closely unless there is plenty of room in which they can move away. When they are in difficulties they will turn and fight savagely with the sharp nails of both feet and are extremely dangerous. A kangaroo can rip open a dog with one stroke of a nail. Because of the early importation of the dingo, kangaroos have acquired an effective technique for dealing with attacks by dogs. If near a waterhole a kangaroo may entice the dog into the water, then hold it underneath the surface until it drowns. Wallaroos are really rock kangaroos and are not quite as large as the other kangaroos. They are dark in colour, savage in temperament, shaggy coated, and not as elegant in appearance as the other members of the family.

K OALAS AND WOMBATS are believed by some biologists to come from an original common stock which lived in tree hollows at ground level. In a distant age of evolution, koala and wombat took opposite directions to protect themselves from the roaming predators of the time. The koala, handicapped by its sluggishness, took to the trees, while the wombat dug his burrow down under the earth. Both are marsupials. The food of the koala is limited to about a dozen varieties of eucalypt, all of which have a high oil content, and which include the manna gum, the blue gum and the red gum. Until the koala became a protected animal quite late in this century, millions were hunted, if that is an appropriate word for the pursuit of so helpless an animal, massacred, and the skins exported all over the world. As late as 1927 Queensland exported 600 000 skins. When the outcry against the slaughter of koalas first started, the skin exporters quickly changed the name to wombat fur.

P OSSUMS come in a great variety of sizes and shapes. At the large end of the range is the foam-plastic-masticating Tasmanian brushtail possum and at the small end of the range the honey possum, a variety that is small enough to feed on some plant flowers like a four-legged terrestrial honeyeater bird. The commonest of the mainland possums is the brushtail, a marsupial tree dweller as common in cities as in the bush, with a very strong territorial sense. Possums exist in great numbers and have been a source of fur since before the turn of the century. A small number of possums introduced into New Zealand multiplied in a few years into an estimated staggering twenty-three million. Generally, too close contact is not advisable. Possums have sharp teeth and a terribly tenacious grip with the jaws once they have bitten. There have been cases where only the administration of a strong anaesthetic would make a possum let go of the feeding hand it had so eagerly snapped up.

There are striped possums, ringtailed possums, pigmy possums, scaly-tailed possums, Leadbeater's possums — believed to be extinct until rediscovered not far from Melbourne in 1961 — and many others including the glider possums whose jumps are turned into short glides by the web membranes between their limbs; a long feathery tail serves as a rudimentary rudder. There are pigmy gliders, sugar gliders, yellow-bellied gliders, greater gliders, squirrel gliders and others. Gliders, like the other possums, are night creatures feeding mainly on insects and the nectar from the flowers of shrubs and eucalypts.

B ANDICOOTS are sharp-nosed little creatures somewhere between the size of a rat and a rabbit and almost invariably the hosts to numbers of ticks and lice. They live on a mixed diet of herbage and mice, lizards, young rats and insects. In order to get at worms and grubs they dig narrow conical pits into which they can insert their long conical noses. The grubs they feed on are usually harmful to cultivation. There are only a few varieties of bandicoots and three, the barred-back bandicoots, the pig-footed bandicoots, and the rabbit bandicoots or bilbies have been brought to extinction or the edge of extinction by the introduction of the fox and the ravages of settlement and bushfires. Bandicoots are, in general, beautiful, useful, harmless creatures with a distinctive charm and a bright good nature, though they will never really be popular with gardeners who do not like conical holes in their lawns and gardens.

MARSUPIAL MICE are amongst the most delightful of all the furry animals of Australia. They are sprightly and intelligent night animals, eating insects and flesh mostly, and endowed with a shining small courage that enables them to take on any animal their own size. Marsupial mice occur in many varieties up and down the continent and they are a sub-family of the family which includes the native cats. The oddest of the marsupial mice (known technically as *Phascogale*) is the brushtailed phascogale which is found from Queensland down into Victoria in sclerophyll forest and woodland, and grows up to 44 centimetres in length. The brushtailed phascogale has been known to attack domestic poultry, but is valued because it preys on mice and rats! Its main diet is insects. As might be expected, the eastern native cat, now almost extinct on the mainland but common in Tasmania still, and the northern native cat, both look disconcertingly like very large mice. In fact the native cats are not really cats at all but marsupials having something in common with the American opossum and the weasel, and measuring about 75 centimetres from nose to tail tip, or only just over half the size of its relation, the very much more aggressive tiger cat which measures about 1.2 metres in length overall. The eastern native cat is carnivorous and is quick to acquire a taste for domestic poultry, though it lives mostly on small creatures like mice, insects and small reptiles. The tiger cat is a much fiercer animal and a skilful tree climber which can hold its own against large domestic cats and small terrier dogs. Despite their ferocity the young, when reared in captivity, are capable of unusual affection for human beings. Because of their habits the tiger cats are in danger of ultimate extinction.

THERE ARE A NUMBER OF RAT-KANGAROOS, which are not really rats at all, but marsupials sharing features of the possum and the kangaroo families. They are usually insect eaters and can be found in desert areas, grassland, and in bush and woodland. The brushtailed rat kangaroo, once common all over the south of Australia but now present only in the south-west of Western Australia, has the pleasing habit of carrying its nesting material coiled up in its tail.

THE TASMANIAN DEVIL is the most formidable animal in Tasmania's bestiary. The Tasmanian devil is believed to have got its name from its black coat, which roused superstitious fears in the early settlers. More heavily made at shoulders than at rump, about 12 kilograms in weight, 30 centimetres high at shoulder, powerfully savage-jawed with strong bone-cruncher teeth, the devil is a scavenger and not as black as he appears from a distance, his coat carrying small patches of white fore and aft. Only able occasionally to hunt down small creatures because of his slow movement, and then only by using exceptional stealth and cunning, the devil lives mostly through his nose, which can detect the smell of carrion from a considerable distance. Devils fight viciously over carrion and are usually badly scarred from their encounters. The noise of a fight of devils is an awe-inspiring mixture of grunts, growls, screams and snarls and as likely a source of their name as their funereal coats, or fierce expressions. The devil is a marsupial and dog-like in many ways, particularly about the skull and in its long canine teeth which give it a tenacious, often lethal grip.

THE TASMANIAN TIGER is sometimes called a wolf, and a hyena, and is neither of these animals. Technically it is known as a thylacine and though very dog-like to look at, is in fact related to the wild cat family and to the Tasmanian devil. Like the questing beast it is an animal of strangely disconnected parts. The head is wolf-like, the body dog-like, the thirteen or so stripes along its back are tiger-like, the tail is kangaroo-like and the female's pouch is the same as that of the Tasmanian devil. Thylacine, as a name, is scientifically precise but a bit abstract and bloodless. Tiger at least suggests ready identification in the bush and is more appropriate for a meat eater standing about 60 centimetres at the shoulder with a lower jaw so articulated to open widely as to make the head almost disappear behind its chasmal gape. The tiger grows up to about 2 metres in length. Any animal that can crush a dog's skull in one slashing bite deserves a reasonably appropriate name, and tiger is at least suggestive of the thylacine's character and ferocity.

The tiger roamed the mainland less than a million years ago and his disappearance from the continent but persistence in Tasmania is a mystery; there is nothing to connect his extinction on the mainland with the presence of the first Aborigines. In Tasmania the tiger's haunts were the hill forests and it quickly became unpopular with the early settlers for its savage appearance and its occasional attacks on sheep. The government of the day paid the high price of one pound sterling for every dead tiger. As early as 1823 the Tasmanian tiger had become a rarity hunted almost into extinction on doubtful evidence and the greed and prejudice of pioneer farmers. The tigers were not usually sheep killers and probably killed only when forced to by sickness, or conditions of bad weather and famine. In a classic case of locking the cage after the tiger had gone, the government did not declare it a protected animal until 1938. Fortunately, the wild and inaccessible ruggedness of much of Tasmania's bush provided a sanctuary for the few survivors, and there have been a number of reliable reports in the early 1970s of its reappearance after decades when the only clue to its survival was the finding of an occasional set of pad prints.

The facts about the tiger's habits are very few. It hunts alone, and is sometimes paired with a female. It eats kangaroo, wallaby, and sheep if in difficulties. The tiger is a highly discriminating eater, drinking the blood of its prey and eating some of the nasal tissues and the liver and kidney fat. It is not believed to eat much, if any, of the meat.

THE DINGO, introduced many centuries ago by the Aborigines, has been suggested as a prime cause of the disappearance of the Tasmanian tiger from the mainland; but this seems unlikely in the light of the evidence of the tiger's capacity for dealing with up to four domestic dogs at a time, and his superior strength and weaponry. On the other hand the dingo was never introduced into Tasmania. A more likely suggestion is that the team of Aborigines and dingo dogs working together exterminated the mainland tigers as they are believed to have exterminated the diprotodon, a giant wombat-like creature, and the giant kangaroo.

Dingoes are widespread through the bush and woodlands of the continent as well as in the open plains country. They are usually of medium dog size, about 60 centimetres at the shoulder, but produce occasional giant specimens measuring up to 2 metres overall. In 1963 one large dingo was mistaken for an escaped circus lion and shot. Dingoes are handsome creatures, mainly tawny yellow in colour, and looking like a smaller variety of wolf. They are found mostly in northern areas now and are a pest detested by farmers. A dingo will kill up to sixty sheep in a night. They have proved an expensive nuisance and an indirect threat to other wildlife. The cost of installing dog-proof fences has been great and the bounty paid for dingo scalps has produced the Australian outback character 'the dogger', a professional trapper. Otherwise dingoes are killed by poison baits which are often laid indiscriminately from the air and which inadvertently cause the destruction of many valuable, useful, and sometimes scarce creatures.

THE SPINY ANT-EATER AND THE PLATYPUS are the oddest of all Australian oddities. They both defy the conventions of all the other mammals by laying eggs but then suckling their young. To complicate matters further the spiny ant-eaters are known by the Latin name *echidna,* which is the name of a genus of eels. Although the echidna looks something like an American porcupine or

European hedgehog it is related to neither, belonging correctly to a genus of its own with the splendid sounding name of *Tachyglossus*. The spines that cover all but the underside of its body are an effective defence against its enemies. In an emergency it sinks slowly into the ground. This is achieved by digging rapidly downwards with its clawed paws, under cover of its spiky all-round umbrella.

While the echidna was something of a perplexing puzzle to early naturalists, the platypus was a shock from which they took a long time to recover: an animal with webbed paws, a broad flat bill like a duck's, a furry mammalian coat, and amphibious habits! The platypus was first sighted in 1797 and controversy about its nature, status, and egg-laying capacity went on for almost a century until 1884. It is not generally known that the platypus carries on its hind limb a claw-shaped spur connected to a reservoir of poison. Biologists have seen this as evidence of a far distant reptile ancestry. This is supported by some other bones of the platypus which occur only in that form in reptiles. The platypus belongs to a quite narrow strip of the coastal border of eastern Australia. Within that strip, which extends along the whole length of the continent, it occupies streams and creeks of every sort from the turbid warmth of tropical waters to the freezing waters of rapid running streams up to 2000 metres high in the Kosciusko alpine area. The platypus lives, with others, in burrows up to 18 metres in length dug in a river bank.

The only fossil evidence of either ant-eater or platypus exists in Australia and is dated back to about a million years ago. There is no trace of them whatever anywhere else in the world.

G ECKOES, like the other lizards, are rough, beady-eyed, warty-looking small creatures of the night. Their skins are scaly, their eyes lidless and in some varieties the skin of the underpart is so thin that the private intestinal workings can be seen. Geckoes have only two sorts of footwear; clawed on the groundlings, and adhesive-padded on the climbers. The pads are strong enough in one common species, the dtella, to enable it to run upside down across ceilings.

The tail of the gecko is a strange one. It comes in a variety of shapes, some imitative of the local vegetation. The southern leaf-tailed gecko has a tail uncannily shaped exactly like the narrow eucalypt leaf. Knob-tailed geckoes accumulate fat in their tails as an insurance against bad times, and have been known to survive for as much as a year on their stores, as long as water was available. Many geckoes have expendable diversionary tails. When attacked they discard their tail. The tail continues to wriggle vigorously, thus acting as a cautionary tail, while they make good their escape. It takes several months to grow a new tail. Camouflage is an important part of the gecko's defences. Moving from light to dark, or dusk to light background, the gecko quickly changes its basic colour to match. In its natural environment it is perfectly camouflaged and reveals itself usually by movement. Geckoes grow to between 8 and 15 centimetres long, and feed mostly on spiders and insects, though cannibalism is not unusual. Having taken an insect in its jaws, the gecko will sometimes bash it repeatedly against rock or tree trunk until it loses consciousness and becomes more amenable to consumption.

Geckoes have defences quite as dramatic as their distracting tails. Some wail like babies, others make yapping, barking noises, and the aural assault is accompanied by an inflation in which the creature opens its mouth wide, rises ferociously on all its legs, and puffs itself out while advancing on its enemy, which is often fooled into flight by the display. Geckoes are completely harmless to larger creatures. They are found throughout the whole of the country.

GOANNAS, a name believed to be a corruption of iguana, are large reptiles very different from the little geckoes. They have large, sharp teeth angled slightly backward, which enable them to grip their prey very efficiently. They can run, often at high speeds, on all four legs, or on the hind two, when the body is reared up almost vertically. Cornered, the goanna rears up, inflates, and hisses loudly. Attacked, it fights back with its heavy tail, which can throw a man off balance.

The lace monitor grows to 2 metres in length and is common in eastern Australia. A tree climber, it will, if startled, race up any tall stationary object in sight, including a human being! The other goanna found throughout the east is the Gould's monitor, a very fast-moving ground dweller which lives for the most part in old rabbit warrens or hollow logs. Most of the other goannas are restricted to the more arid inland areas of the continent.

SNAKES have always had a profound psychological effect on people and have served as important symbols down the ages. Fear of, or liking and reverence for, snakes is a component of the character of most human beings and therefore both are natural reactions.

Australia has about one hundred and thirty species of snakes, of which about seventy-five, or more than half, are venomous. Fortunately venomous does not necessarily mean dangerous to human beings. Perhaps a dozen snakes present a real threat and these are dealt with in Chapter 15.

There are about twenty snakes which are completely harmless to man. These are the burrowing worm snakes which are blind, and are found throughout Australia except in Tasmania. These snakes are small, rarely more than 30 centimetres in length, and look more like large shiny scaled worms than snakes. Their mouths are so tiny that it is impossible for them to bite any part of the human body. Their normal diet is beetles, bugs and insects. They are not venomous and live below ground much of the time.

The ten varieties of python found in Australia are also considered to be harmless to man, despite their reputation for ferocity and their fictional image as crushing machines forever seeking human prey. Some can grow to a great size, 3 to 8 metres in length, and are capable of astonishing swallowing feats. A famous photograph shows a python with a huge bulge in its middle. The bulge is a whole wallaby which, after being squeezed to death, was swallowed whole through the snake's very loosely articulated jaws. Despite their lack of venom and harmlessness to man, pythons are best kept at a safe respectful distance. The best known of the pythons is the carpet snake which can grow up to 4 metres long, and is common everywhere except in the south-east of the continent.

Beyond the blind snakes and the pythons there are about a hundred species, most of which are harmless but which are difficult to identify and separate from the dangerous varieties. Because of this *all* snakes should be treated with the utmost caution. Many varieties are venomous but harmless to human beings because their venom does not work on mammals, or is too weak to do damage, or they are so constructed that they cannot inject enough of their venom. Others, like the blind snakes, are too small to get a grip, or the venom glands produce only enough poison to deal with very small creatures. One apparently highly discriminating group simply refuses to bite human beings. An explanation that has been suggested is that the odour of human beings is repugnant to them. A more likely explanation is that human flesh is as venomous to them as their venom is to us.

Lizards and snakes are sometimes hard to identify since some lizards look like, and share the colouring of, snakes. Also, some lizards have only tiny vestigial legs which are difficult to see. The most distinctive identifying feature of the lizards is their tail. Snakes are not really fast movers, though in flight they can give an impression of high speed. A sedate 6 or 7 kilometres per hour is the best that they can achieve. Snakes swallow their victims whole and are provided with very strong, quick-acting digestive juices which help the prey through the system quite speedily. The digestive juices work properly only at warm temperatures, so snakes feed only when it is warm. A cold snap, catching a snake with its prey only half digested, stops the action of the juices. The prey begins to rot, fermentation occurs, and the snake may die of poisoning from its own victim.

The venom in snakes has only one purpose: to subdue and usually kill the prey so that it may be more easily and conveniently ingested. The snake bites suffered by human beings are therefore attempts at defence by the snake who must have felt threatened. A sharp lookout and deliberate movement are the best ways of avoiding encounters with snakes. Haste and carelessness can lead into serious danger. A snake, trodden on or kicked, or disturbed unexpectedly is bound to feel threatened and will attack immediately. Many snakes live in trees and it is a good idea, if trunks and branches are being used as supports or to help progress, to look where hands are being put or bodies rested. Thick grass and undergrowth should always be treated with care. Putting hands down rabbit holes, into hollow trees or logs, or under logs, without being able to see what they contain or conceal is to ask for certain, and perhaps lethal, trouble.

SEA SNAKES are most common in northern tropical waters, but are found quite commonly as far south as Tasmania. The yellow-bellied sea snake, which is often found washed up on the east coast, especially in New South Wales, is the most widespread of the sea snakes, being found all over the Pacific and Indian oceans. Sea snakes are venomous but their fangs are small and their jaws quite inefficient. No case of death from sea snake bite had ever been recorded in Australia until the late 1960s. The sea snakes live mostly on fish and are seldom aggressive in their behaviour. Their tails are flat and broad to enable them to move easily through water. One group of sea snakes, a small one, is confined to coastal mangrove swamps or coral reefs. These snakes can move about easily on land as well as in the sea and are often capable of climbing trees. It is believed that some sea snakes live solely on a diet of eels.

THE MULTI-FACETED INSECT MICRO-LIFE of the bush holds about a million different varieties. Like so much bush life, the insects are unassuming, unobtrusive, self-concealing to the vague gaze of the urban explorer who knows neither where to look nor what to look for, but who suddenly develops unwelcome eyes, of a sort, in the skin, when red bulldog ants make a sudden, stinging attack.

ANTS have unbelievably complex forms of social organisation with features bizarre enough for the most outlandish science fiction. There are over 1000 varieties in Australia, one of which is the honeypot ant. This variety uses a specialised form of worker ant in which to store food. This ant carries a sac that is filled by the other ants with food to many times its own size until it is totally immobilised. When the other ants need food it is delivered as requested. The storage ant is eventually emptied and then resumes the normal tenor of its way, presumably with a sigh of relief. Most ant colonies have a distinct class structure of males, females, and (sterile) workers presided over by a queen. Their habits are ruthlessly unpleasant. Mating occurs in the air after which the partners fall to the ground and the male dies. The females produce young which are fed by the queen who, immobilised in her task, turns her wing muscles into nutriment.

WASPS are provided with stings in order, like snakes, to paralyse their prey, not especially to harass man. Having subdued a victim the wasp lays its eggs on the recumbent body. When the young hatch out they feed on the body. Wasps have many specialised forms. In the flower wasps, the females are wingless and the males wings. The mason wasps specialise in the construction of nests made of dried mud where food is stored. The cicada hunter wasps are large orange and black wasps which live in earth burrows where they store the paralysed cicadas they catch.

BEES belong to the same family as ants and wasps, and are very like wasps to look at but do not have a highly organised social structure like the ants, or like the minority species the hive bee.

DRAGONFLIES AND DAMSELFLIES are amongst the largest and most exquisitely delicate of the insects. Their colours are brilliantly blended, their large, transparent, veined wings dazzling in sunshine, their capacity for involved aerobatics a revelation and a delight. These insects have very large globular eyes with sharp sight of a high order. Their flying power and speed enable them to catch the flies which are their prey while in flight. With a wingspan of up to 12 centimetres the dragon and damselflies are the most obvious of Australian insects, particularly near streams and dams.

BUTTERFLIES AND MOTHS are common from the tropical bush of the far north to the temperate rainforests of Tasmania. The largest of the tropical butterflies is the Cairns birdwing female which has a wingspan of 21 centimetres. The male grows to 16 centimetres in width. There are a number of Australian butterfly families, some of which, like the imperial white (the wings are white underneath) and the wanderer or monarch, are brilliantly coloured. The families include the whites, wanderers, blues and coppers, nymphs, swallowtails, browns and skippers.

Australian moths are often almost as colourful as butterflies and there are several thousand native species as well as some hundred introduced from overseas. In the larval (grub) stage moths are capable of extensive damage to crops, and in some cases, stored food. Cutworms, which bite through the stems of plants, are the larval stage of the cutworm moth. The emperor moths which feed on the leaves of pepper trees and eucalypts are amongst the largest of the moths with a wingspan of up to 9 centimetres.

FLIES are ugly, omnipresent, harmful, dangerously insanitary, and in the form of the mosquito can be lethal if carrying any one of a number of fevers. The Upper Murray area has been the focus of mosquito-carried encephalitis outbreaks which have recently been recurring every few years. It is the female which feeds by sucking blood and which transmits the diseases. Although there are over a hundred species of mosquito in Australia they are more of an irritating nuisance than anything else.

MARCH FLIES are large, with strong beaks which can inflict a large lumpy painful bite. Their relations run into hundreds and they can make life uncomfortable anywhere in Australia in one form or another. Some people react to bites by coming up in large lumps which have to be treated with anti-allergy compounds by a doctor. A few of the flies that menace the country are the housefly (disease spreader, bringer of typhoid, dysentery, and many other diseases), the fruitfly and the blowfly. The largest is the assassin fly which has a wingspan of up to 9 centimetres and which catches its prey in flight like the dragon and damselflies.

A good but rather sticky fly repellent which keeps off all but the most insanely aggressive of the stinging/biting flies can be made up from equal quantities of olive oil and the antiseptic Dettol. Shake before use and apply liberally. No one will come near you afterwards but then neither will the flies. The remedy is not advised for people with sensitive skins. This is a New Zealand unguent and works, usually, against the New Zealand sandflies which are a gaunt and savage breed of their own equalled only by their relatives in the Cairns area of Queensland and further north.

SPIDERS differ from insects for a number of reasons, two of which are that they have eight jointed legs, and never have wings as some insects do. Spiders, like snakes, seem to trigger off deep unconscious fears and loathings in many people, making their worst nightmares into intolerable reality, a reaction prompted more by the spiders' often hideous appearance than by their actual lethal potential. A few spiders are very dangerous; most are not. The dangerous ones are dealt with in Chapter 15. Again, as with snakes, identification is often difficult and it is best usually to treat all spiders with respect.

Despite their nasty looks and occasionally lethal potential, spiders, which occur in uncountable varieties in Australia, are fascinatingly gifted creatures, particularly in their ability to spin fine strong thread which is used to build the familiar wheel-shaped webs we know in gardens and in the bush, but also sheet webs, lattice webs, funnel webs, and net webs which are cast over the prey entangling the victim until the spider can move in and kill it. Some Australian varieties are the jumping spiders, night hunters, wolf spiders, lynx spiders, flower spiders, and huntsman spiders.

SCORPIONS belong to the spider family whose relatives include the centipedes and crabs. There are only two varieties of true scorpion in Australia and these are found all over the continent. They are small in size in the southern half but grow larger in the tropics. The poisonousness of the sting in the tail of the scorpion varies according to the size of the creature. Tropical specimens can give a dangerously poisonous sting, but the amount of venom in Australian scorpions is rarely enough to endanger adults.

TICKS AND MITES also belong to the spider family. Both are skin parasites, but mites are minute while ticks may be as big as half a centimetre. Ticks are dangerous partly because of the diseases they can transmit to stock and to human beings and partly because of the poisons they inject which can ultimately result in paralysis. Some people are sensitive to tick venom and suffer a severe allergic reaction. Most ticks are blood-sucking parasites of the skin of a number of animals. Some of the more common Australian ticks are the scrub tick, the cattle tick, the fowl tick, and the brown dog tick. Ticks are coarse feeders, gorging themselves on blood to several times their normal size until, with their heads deep into the flesh, they begin to look more like skin eruptions than the dangerous parasites they are. The ticks are just as likely to take up residence in, and on, the skin of human beings as on the skin of an animal. How best to deal with ticks is covered in Chapter 15.

The bush teems with life, from tropical swamps to snowy highlands. The Australian natural environment, particularly in the east and south-east, is one of the richest in the world and far beyond the scope of a single, short chapter, or even a quite lengthy book. Although the bush has changed, although thanks to the heedless, mindless greed of its industrial exploiters and the determined ignorance of its agricultural exploiters it is now only a pale reflection of what it once was, it is still the greatest natural resource freely available to all Australians. The more we know, understand and sympathise with the bush and its unique and wonderful patterns of life, the greater the chance of saving it from eventual total destruction ground between the slow spreading natural deserts of the centre of the continent and the fast spreading man-made totally destructive city deserts of the coastline.

Further Reading

Golden Press, *The Gould League Book of Australian Endangered Wildlife*, 1983.
M. Senior, *Australian Animals*, Kangaroo Press, 1987.
I. G. Read, *The Bush*, Reed, 1987.
G. Pizzey, *A Separate Creation*, Curry O'Neil, 1985.
N. Drayson, *Wildlife*, Collins, 1988.
D. Newman, *Australia's Dangerous Wildlife*, Child and Associates, 1989.
S. K. Sutherland, *Australian Animal Toxins*, Oxford University Press, 1985.

CHAPTER 12

GETTING LOST AND OTHER EMERGENCIES

We have erred and strayed from thy ways like lost sheep.
THE BOOK OF COMMON PRAYER

A schoolgirl answered the question 'In what countries are elephants found?' 'Elephants are very large and intelligent animals and are seldom lost.'
JAMES AGATE

Getting temporarily lost while still in possession of map, watch, compass, supplies and equipment is not a very serious matter and may be a most valuable experience. Getting *seriously lost* is a very serious, and possibly deadly matter. To be seriously lost means that all or most of the normal equipment has been left behind, become damaged beyond repair after an accident, been swept away by a river, or otherwise removed and destroyed by external circumstances. Getting seriously lost is a very exceptional circumstance. There is nothing wrong with getting seriously lost. There is everything wrong with getting seriously lost and not being prepared for it with knowledge and the internal resources to deal with the situation.

SURVIVAL is primarily a psychological affair. It is necessary first of all to accept the fact that one is seriously lost. Most people who go unprepared into the bush and get seriously lost are not at first prepared to admit their situation. Action, any action, no matter into what hazard it may carry you, seems preferable to thought and analysis of the situation. If this tendency towards panic action shows itself, it is important to know what you are running away from. Usually it is yourself and your own feeling of inadequacy in these unprecedented circumstances. There are other factors operating as well: there is fear of the unknown, and fear of the suddenly hostile bush. Nature, the bush, the surroundings, are not hostile, they are neutral. It is your hostility that fills the bush with what is not objectively there. The bush is neutral. Therefore, if you take thought, you can probably make use of it. Fear of the unknown does not arise in people who are prepared with knowledge, who have learned about the bush and its resources and how to travel safely in emergencies, and how to provide basics like food and shelter. Knowledge prevents panic, worry, and waste of energy needed to get out of the situation.

The basic worry, usually unacknowledged, is about yourself. Are you adequate to deal with this intimidating and frightening situation? Commonsense is the best reassurance. People who are capable of going into the bush are capable of getting out of it again. Many people have been lost before, some in worse circumstances, and have got out safely. Quite small children have survived in the bush for many days. Of course you are capable of getting out, as long as you keep your head and your wits about you.

If you suspect that you are seriously lost, stop and examine the situation. Don't allow yourself to be driven into possible danger by a doubt you refuse to admit to yourself. You may find, when you sit down, rest, and look calmly at things that you aren't lost at all. This has happened to people! If you find that you are lost, stay resting and examine your resources. Look through all your pockets, think clearly about the type of bush or country you are in and its possibilities. Don't jettison *any* equipment remaining with you. What you carry with you and in your pockets are possibly priceless assets. You are your greatest asset. If you are determined to survive, you will survive.

L OOKING AFTER YOURSELF is the best way to start dealing with being seriously lost. Don't waste energy. Rest whenever you can. Sleep comfortably whenever you can. Try to relax yourself as much as possible. Tension and worry consume prodigious amounts of energy. Make haste *slowly*. If you are on the move, don't rush. Rest regularly for a time in every hour. Don't be afraid to stop and think things through again in a comfortable, relaxed way. Don't move unless, or until, you have a plan to follow. If night is coming on, give your problems away, make camp, be as comfortable as possible, try to get in a good night's sleep.

W ATER IS LIFE. The body can survive without food for days and weeks; without water it is near to death within only a few days. It is important to find water if there is none nearby, and it is essential to drink plenty of it *whenever it is available*. Quite a small loss of water by sweating can begin to seriously affect bodily comfort, clarity of thought, and capacity for decision. Drinking plenty of water is a first priority.

Finding water, unless in a desert, or arid area, is not usually too difficult. Watercourses and soaks are usually fairly obvious and even dried up watercourses almost always have pools of water suitable for drinking, along their lengths. If not, water can be reached by digging along the course. In particularly difficult areas water is indicated in the vicinity by ants, wild bees, mason flies, finches, and wild pigeons. Other birds, insects and mammals are not reliable indicators of water. Ants need water, and if a column of ants is discovered it can

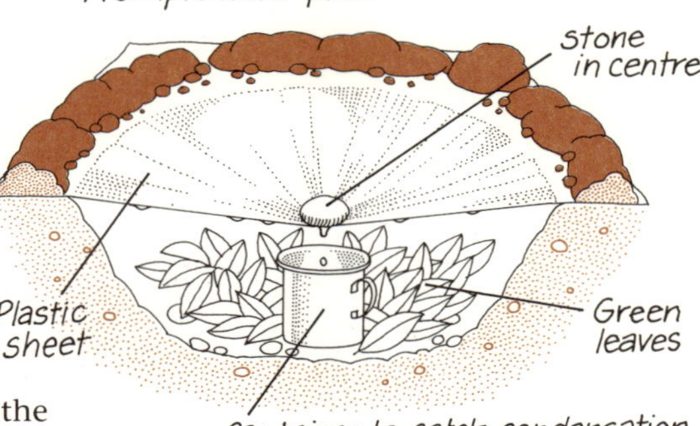

often be followed to a water supply in a hole in a crotch of a tree. This water may be 'mopped' out using grass fixed to the end of a stick, or sucked out through a hollow straw. Pure water should be clear and have no smell. Water purification tablets of the sort described at the end of Chapter 15 should always be carried in one's personal emergency kit.

SHELTER can nearly always be improvised: in the lee of a fallen tree, in a fallen hollow tree (after testing for snakes or other inhabitants with a long stick), in the lee of a rock, or rocky outcrop. If there is rain about, a lean-to of interlaced small branches roughly thatched with leaves,small leafy branches, bark and anything else that might be suitable is quick and easy to build and will keep off most rain if set up in a naturally protected place. Living foliage should only be used in the direst emergency where life is at stake. The bush is a too precious and too fast diminishing resource to be used casually to avoid a shower or two. Bedding is important for a comfortable sleep and to insulate the body from the cold from the ground. Fern leaves, leafy branches, leaves, moss and grass can make a very good bed. If no lean-to has been made extra warmth and protection can come from a covering of foliage. If it is possible to make a fire, this is a cheerful and reassuring sight as well as an extra source of warmth. Accumulate plenty of firewood before going to bed.

FIREMAKING AND FIRELIGHTING depend entirely on dry materials. If it is raining and the obvious materials are damp or wet then they must either be dried or dry materials found. Very small, thin twigs that are only damp can be put next to the body inside one's clothing, while more substantial twigs are divested of their wet outer bark with a knife and cut into dry shavings. Small, almost dry kindling can often be found under fallen trees, and in sheltered places. Dry thin bark, if it can be found, is a big help. The more dry materials that are available the better. It is a waste of time, patience, and valuable matches to attempt to light a fire without an adequate supply of small, fine, *dry* kindling to ensure that it grows and quickly becomes big enough to dry out larger material. Don't attempt to light a fire on wet grass or other herbage. Put stones together to form a fireplace or lay down stripped small branches side-by-side. Before lighting the fire be sure to have wood ready to put around it to dry. Dead standing timber is likely to be drier than

Emergency shelter : a lean-to may be built from material found in the bush.

timber on the ground. Be patient, build up the dry materials for the fire with care. There is no need to split matches to make them go further if your fire materials are abundant and properly prepared: *one* good match can do the job. In dry weather, firemaking and firelighting are usually easy, and the main consideration is to site the fire where it is unlikely to spread to the surrounding bush.

EXPERIENCE is the best teacher. It is worthwhile whenever bushwalking in wet conditions to make a fire for the practice it gives, whether it is needed or not. Restrict aids to a pocketknife and matches. Since excellent waterproof matches are now on sale generally, these are the sort which should always be carried. It is almost impossible to make a fire without matches unless sunlight and a lens are available. The lens, a spectacle lens will do, concentrates the heat of the sun in a tiny white spot of light which can ignite really fine dry timber which can then be gently blown into flame. This is much more difficult than it sounds and *it is easier always to have matches in your pocket.*

HYPOTHERMIA is the most dangerous threat to anyone seriously lost except in the height of the heat of summer, and is still then a possible threat at very high altitudes. In winter, especially as altitude rises, the threat is one that needs to be taken seriously, particularly in the dark. Essentially, hypothermia is a type of exposure involving cold conditions and massive heat loss in the victim. The range of body temperatures at which a human being can live and perform tasks efficiently is a very narrow one of only a few degrees on either side of the normal average of 37° Celsius. As body temperature begins to drop, the work of some of the organs is impaired. The electrical waves of the brain are reduced, the heart rates begins to slow, the liver is affected and with it the supply of glycogen into the blood, the cooling muscles are unable to make use of the blood oxygen and lactic acid builds up and is absorbed into the blood where it can trigger heart failure if the heart temperature falls to 32.22° Celsius or lower. The heart cannot function at all below about 29° Celsius. Lungs and skin stand cooling well, but skin begins to die if it has a temperature just below the freezing point of 0° Celsius.

HEAT LOSS can occur in many ways: through conduction by contact, radiation, convection, wind, rain, water, snow. To lie down on snow, ice, or frozen ground is to invite rapid heat loss through conduction. A number of layers of thin clothing prevent heat loss by radiation better than one or two thick garments. Leakage of heat to the surrounding cold air via convection currents is best stopped by a material which traps air, e.g. wool, which is the best fabric insulator. Wind snatches heat from the body quickly but can be stopped by a windproof outer layer of clothing. Rain and snow which take heat from the body by evaporation, and by directly absorbing and carrying it away, require that outer clothing be *waterproof* (showerproof is completely inadequate), as well as windproof. Cold water promotes heat loss very rapidly. A few minutes' immersion in a stream is equal to many hours of heat loss from cold air and can continue at a high level for long afterwards.

Hypothermia is a condition of massive heat loss usually encountered by outdoor people in very cold conditions accompanied by strong winds with sleet, rain, or snow. Hypothermia can occur at various times of the year throughout much of eastern and south-eastern Australia. Altitude and night increase the possibility. The condition is a very insidious one because of its unobtrusive effects in the early stages. A person in the first stages of hypothermia will become quite incoherent, vague in outlook, suffer from impaired vision, move slowly and ineffectively with unnecessary stumbling and frequent falling down. As the brain cools, the surroundings lose their reality, and indifference and irrationality show themselves. As the muscles cool co-ordination becomes more difficult.

Hypothermia is usually a combination of physical exhaustion, low blood glucose, cooling brain and muscles. It should not be confused with the slowness and lack of co-ordination that occurs from extreme physical tiredness, nor with the irritability, confused outlook, and occasional collapse that come simply from a low level of glucose in the blood caused by self-neglect in not having regular snacks. Both conditions can occur at almost any temperature and are quickly remedied by rest and food. Hypothermia is most likely at low temperatures. The people likely to succumb first to hypothermia are usually young, thin, male, debilitated or unfit, or the injured; but if conditions are bad enough hypothermia is no respecter of persons. If early symptoms are overlooked or ignored, total collapse into an unconscious, or semi-conscious state soon follows. Cooling speeds up, organs begin to fail and after some hours the patient dies. In this last stage convulsions may occur, with vomiting, and there is a great risk that the victim may die from inhaling his own vomit.

Once under way hypothermia is very difficult to treat in the field and the chances of survival are poor. Prevention therefore comes before treatment, and prevention begins with proper clothing. At most times of the year the rucksack should contain windproof and waterproof outer clothing, overpants and parka, spare pullovers made of wool, and a woollen hat. (The head is a major source of heat loss in cold weather: old timers used to say that cold feet were best cured by putting on a hat.) In winter most clothing — trousers, underwear, shirt, gloves, and socks — should be made of wool, which is incomparable insulating material and maintains warmth even when it is soaking wet. Wool resists water, holds in heat, is only partly impaired in its efficiency when wet. Woollen clothing is one of the best preventers of hypothermia. In comparison cotton and synthetics are cold, clammy, quick cooling and therefore dangerous in cold conditions.

Physical fitness is important. It is an invitation to disaster to take along someone who has not quite recovered or is still weak from illness. Young people, thin and still growing, are more liable to hypothermia than most. If conditions look like being rough, cold, and difficult, young people should not be taken. If they are taken, then the distances to be covered have to be substantially reduced to allow for their weakness. Distance, in any case, should always be within the capabilities of the weakest member of the party. The absolute minimum number for any party likely to meet cold and inclement weather is four.

Equipment too is important. Only first class sleeping bags with adequate covers are good enough for bad conditions. Plenty of food should be carried for daily meals and frequent snacks, as well as a substantial quick energy food supply. It is *always* a good thing to leave word behind with a reliable person about where you are going and when you expect to return.

If conditions deteriorate rapidly, never be afraid to turn back to safety. This is the most courageous and often the most difficult thing to do, but is much better than running the risk of the nightmare experience of hypothermia with its too frequent consequence of swift death.

Treatment of a victim in the first stages of hypothermia is straightforward. Mild cases, which can be dealt with effectively, show tiredness, vagueness, and clumsy performance of all but the simplest tasks. At this stage body temperature has not dropped below 35° Celsius and the first need is to stop the accelerating loss of heat. This can be done by getting the patient out of wind, cold and wet into the shelter of a tent, cave, or lean-to. Wet clothing is removed and dry woollen clothing substituted. He is then put into a sleeping bag with a warm member of the party and given food and hot drinks. In this way heat loss is stopped and new generation of warmth is begun. At this point there is still a good chance that normal body temperature can be restored. Alcohol must not be given. *Early recognition of hypothermia is crucial to recovery.*

Below a body temperature of 35° Celsius the situation is extremely dangerous and desperate remedies may have to be undertaken. Below this temperature the outer layers of the body are very cold and the vital core is cooling rapidly. Death may be only a couple of hours away. Heat generation has stopped, body chemistry is unbalanced and even small changes may stop the heart. From here on the body temperature will drop considerably before rewarming starts, despite the most effective measures. At this stage the victim may be incoherent, confused in thought, forgetful, and have muscular rigidity with erratic, jerky movements. Unconsciousness is not far away. Shelter, dry woollen clothing, and a sleeping bag with a warm companion (two people stripped to underclothes: one either side is better if a double sleeping bag is available) are immediate essentials. Hypothermia victims should never be put in cold sleeping bags. They have not the heat necessary to warm the bag or themselves. If the bag can be quickly warmed with rocks, canteens, or bottles so much the better, though the objects used should only be warm, not hot. If possible fires should be built on either side.

Food and drink may now be dangerous since the patient may not be able to swallow without choking and may be unable to digest the food which remains in the stomach as a potential source of vomiting and choking. The body, and the heart in particular, is now at an extremely critical stage. Any movement or exertion by the patient may lead to heart failure. The patient must remain lying down. Sitting up or standing may be fatal until he has completely recovered, which will take a minimum of three days. Even movement by stretcher to better quarters may be fatal. The greatest care must be taken in warming the patient. Close exposure to fires or hot stoves may alter the body's internal balance and bring immediate death. In any case they may cause very deep serious burns.

The first concern of anyone seriously lost in winter must be the prevention of hypothermia before anything else.

FROSTBITE is another winter hazard and much more real than many people realise. Only recently an inadequately clothed midwinter bushwalker suffered from frostbitten legs during a Sunday morning stroll through hilly bush only about 80 kilometres from Melbourne. Frostbite occurs when tissues lose heat faster than blood circulation can replace it. High winds make frostbite much more likely. The face, fingers and toes are the most common areas affected. Treatment is largely a matter of *dont's*. Warming by contact with warm skin is permissible, but not much else. Hands can be warmed under one's own armpits, feet and face on the warm skin of another member of the party. *Don't* rub snow on an area suspected of frostbite, it will only make the damage worse. *Don't* warm extremities by holding them close to a fire. If the frostbite is severe, the area having no feeling, the skin white, and the flesh deeply frozen, some thawing can be achieved by immersing in water *below* 40° Celsius and above 37.2° Celsius. Often it is better to wait for medical help rather than risk a partial thawing and then a further freezing. Frozen feet should be left alone for medical treatment. Frozen, they may be used for walking without much increasing damage. Thawed, they can cause intense pain and the victim has to be carried.

HEAT EXHAUSTION happens when exercise, effort or very hot conditions cause dilation of the skin's blood vessels drawing away blood from the brain and starving other organs of the body. A feeling of faintness follows, which may be accompanied by weakness, a racing pulse, nausea, cold clamminess of the skin. Salt deficiency in the body and minor degree of dehydration can continue to heat exhaustion. Treatment involves replacing body liquids by frequent drinks, replacing body salts with salt tablets or a weak salt solution, and prolonged rest preferably out of direct sunlight.

HEAT STROKE is a much more serious matter. Symptoms show themselves in a flushed hot face, pronounced racing pulse, head pains, dizziness and weakness. Treatment, which should be carried out as quickly as possible, is by cooling the head and body with water, or snow, and vigorous fanning. Cold drinks should be taken until body temperature drops to normal.

EMERGENCY NAVIGATION. No one venturing into the bush can consider himself properly trained and prepared until able to orientate himself roughly in relation to the sun and the stars without map or compass. Navigation, the ability to find one's way safely and reasonably accurately, is the foundation of any relationship with the environment. In an emergency the invisible navigation equipment carried in one's head may ensure one's survival.

The group of stars called *Crux,* or the Southern Cross, can be

Navigation by Stars

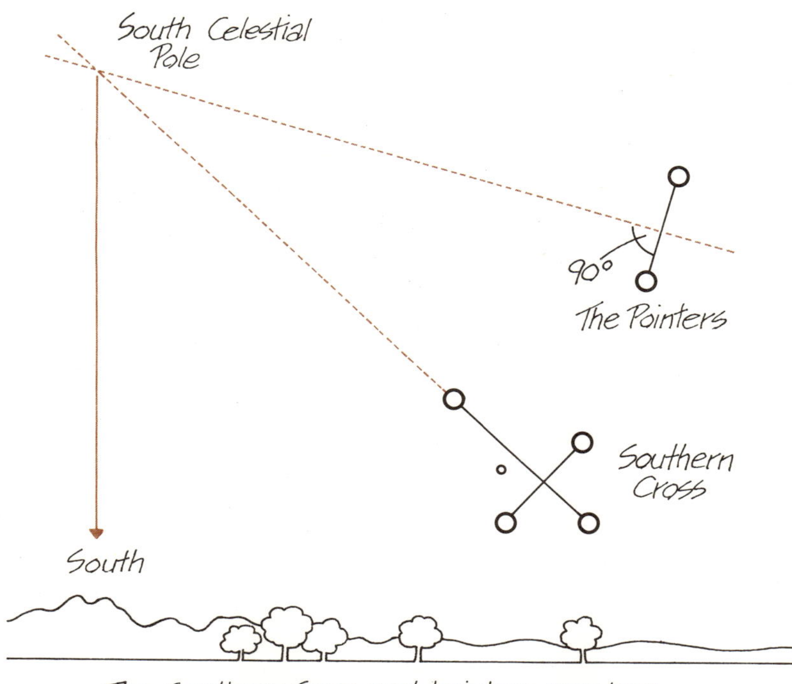

The Southern Cross and pointers may be a different way up at any time of night but the method remains the same.

seen low on the horizon on any clear night. There are at least two similar star crosses in the same area, so care should be exercised when making the identification. The Southern Cross has five stars, the extra one being just inside the cross and below the crossbar. Close by and to the east of the Southern Cross are two bright stars called the Pointers. South is quickly found by extending, in the mind's eye, the long axis of the Southern Cross beyond its foot for about four-and-a-half times its length. A line dropped to the horizon from this point gives due south. South can also be found by extending a line from midway between the Pointers in the same direction as the foot of the Cross. Where the imaginary lines from Cross and Pointers meet, a line dropped to the horizon gives south.

The sun rises in the east and sets in the west. The stars travel from east to west. At its highest point in the sky, in the southern hemisphere, the sun will be due north. These simple facts, unknown to, or forgotten by, most urban people, can be life savers in an emergency. A rough but adequate direction may be found by taking a straight twig about the thickness of a pencil, sharpening one end if necessary, and sticking it in the ground top end towards the sun, so that little or no shadow is cast. After between fifteen to thirty minutes a shadow *will* be cast and it will lie roughly east-west. In the southern hemisphere north can then be found at approximately a right angle to this east-west line. The end of the line that is east or west is easily found by the position of the sun in the sky, the sun being in the east in the morning and the west in the afternoon. This method is the simplest, the easiest to remember, the least time-wasting, and therefore a good one to use in an emergency situation.

When you face north, the east is to your right, west to your left and south at your back. From north in a clockwise direction there are 90 degrees to east, a further 90 degrees to south, another 90 degrees to west, and a final 90 degrees back to north, making 360 degrees in all.

These bits of information are basic. They should be memorised and used as often as possible until they are unforgettable and their use automatic even in everyday situations in town and city. In an emergency, if you are seriously lost, knowing the direction to travel towards safety is the first need. There are many other ways of finding direction and more accurately; many of them are novel, enjoyable and interesting in themselves, but they are usually time-consuming and quite complicated to arrange. Precise navigation is not what is needed in this sort of situation and time is usually precious. Knowing the approximate direction of safety and succour and being able to find it without delay is what is important.

If lost from a camp, which is off a track along a high ridge, the essential thing is to be able to regain the ridge, not to try to navigate exactly to the campsite. Once on the ridge with its track, safety has been achieved and the camp is quickly found. In an emergency situation of this sort it is usually best to select a prominent feature of the countryside (track, river, ridge) in the right direction, and make towards it. This is why it is always so important to know from your map where you are, and the most prominent features of the surrounding landscape. Whatever may then happen to the map it is possible to orientate yourself in relation to the lie of the land and start moving in the general direction of safety along a suitable memorised feature.

CONSIDER EVERY MOVE CAREFULLY. There is no point in moving for the sake of moving; it may actually worsen the situation. If enough time has passed for search parties to be out it is best to stay in one place making signals by whistle (regular double blasts in groups of three), shouting at regular intervals or making smoke if this is possible and is a safe thing to do. Given time, lost people are usually found.

If you are quite sure of the direction to take and it is safe and sensible to do so, then begin moving. *Don't* wander aimlessly, wasting time and energy.

Most people tend to move in circles unless they have aligned their route with a landmark. The reason for this is that the human body is asymmetrical and one leg is invariably shorter than the other. There is no general rule that people who are right handed move to the right and vice-versa. Veering increases the faster you move. It is easy to find out the direction of one's personal deviation. A blindfold walk on snow or an empty sandy beach, or in an open area with someone to watch for deviation will establish the personal tendency in a few minutes. Left or right deviation can cause trouble if it is not allowed for in mist or fog, bad visibility, and at night. In normal daytime conditions this can be prevented by always aiming at a definite point, a landmark, or in thick bush a tree perhaps not many yards away. From the point you have reached select another on the same line and move on to that. A straight line is best kept by lining up landmarks in a row all the time you are moving. It is a good idea to mark the direction you are taking by marking trees behind you or breaking branches and pointing them in the direction you are taking, or arranging stones on the ground, though these can be difficult to see from more than a few metres away. If you come up against an impassable obstacle you may want to find your way back, and if, despite everything, you are walking in circles this will help you to find out the fact.

The basics of navigation and pathfinding are essentially simple. Keep emergency information clear, concise, and concentrated. It is then readily available and more easily and correctly remembered when it is needed. If in doubt make up a small reminder card — to carry with your personal emergency kit. The card should contain the navigation material in this chapter rendered down into a short form that is most easily and quickly understood by you. It is better to have your own brief summary because it is easier and quicker to absorb, and should create no ambiguities or doubts. Besides, what you have prepared for yourself is a better, more confidence-giving psychological support than anything prepared by someone else.

EMERGENCY FOODS in concentrated form are easily found and should always be carried. But don't worry too much about food. Remind yourself that you can carry on for many days without it as long as you can find liquid to drink. It is a good idea to always carry in your pockets some boiled sweets (fruit drops, barley sugar) and some glucose or Ovaltine tablets for quick energy release. Vegetarian and health food shops have a wide variety of dried fruit and nut blocks. These are particularly palatable and nourishing. In the early stages of an emergency, food is probably more of a psychological and morale booster than anything else and the considerable value of the lift it gives has to be weighed against the necessity for strict rationing if it is going to take a long time to get out or to be found.

FINDING FOOD may be difficult, and may not be worth the energy, which would be better conserved or spent in some more useful way. For the Aborigines finding food was an occupation that took up a lot of time, and for a white person without their traditions and skills in recognising and discovering native foods it will obviously be very much more difficult. If food becomes a dire necessity then it is best to concentrate on a few sorts, the most likely to be available and the most easily found.

FISH are a valuable food. Creeks, streams and rivers are usually well stocked with trout, redfin, blackfish, mullet, tench and catfish. All are edible and spread throughout the streams and rivers of eastern Australia. Most can be caught on a bait of worms, grubs, beetles or yabbies. Every personal emergency kit should contain a few hooks and a length of fishing nylon of about 10 or 15 metres and 6 kilometres breaking strain.

Fishing with rudimentary tackle like this is usually futile in the hours of daylight. The hook, line and bait are best set in the evening and left overnight. Deep pools, and deep stretches of water are usually the best places for a nightline. Deep is here a relative term. So long as the

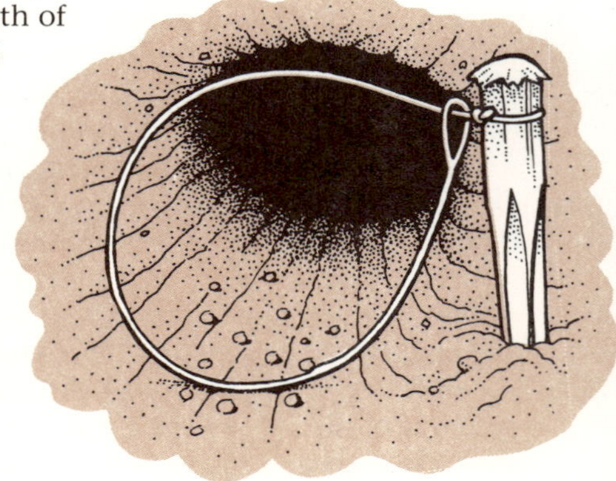

Snare set over rabbit burrow.

138

pool or stretch is deep in comparison with the rest of the water it will allow concealment and hiding places for the best fish and will be water they must pass through when feeding at night. The line should be tied to a strong bush or tree, or a stake pushed deep into the ground and the baited hook thrown into the deepest part of the water. No weight is needed. The bait should be put firmly on the hook, and the hook forced several times through the body. It will then quickly become waterlogged and sink to the bottom of its own accord.

It is very important to allow out only the right amount of line. Too much will let the fish swim away with the bait and entangle the line on underwater snags which can make the retrieval of the fish impossible. Too little will leave no slack in the line and warn the fish off by the bait's resistance to movement. When the line is set, *go away and leave it alone until morning.* Get well away where movement or vibration cannot be picked up by the fish. Resist all temptation to go and see if it has caught anything. Do not interfere. If left to its leisurely self, the fish will get hooked of its own volition. If you try to help by interfering it will almost certainly get away.

There are many advantages to nightlining for anyone who is seriously lost and has to keep body and soul together until found. The equipment is very light, and bulk is negligible. No great effort is needed and rest can be taken while the fish does the work.

FLESH is probably best got from lizards, goannas and snakes. These reptiles are quite common and their flesh is perfectly edible, adequately nourishing, and a good source of protein. All have clean white flesh that is safe to eat once intestines have been cut out. Snakes are only an exception if, in the course of catching and killing the snake, it has bitten itself, when it immediately becomes unsafe to eat. A snake you have not killed yourself is obviously unsafe to eat. Cooking can be done in hot ashes and takes about half-an-hour. Reptiles are cooked with the skin left on and the resulting meat is white, and tastes something like chicken. If there are rabbits about it may be worthwhile sacrificing a short piece of nylon fishing line in order to improvise a snare. When the snare is set rub it carefully all over with soil to get rid of any human scent that may still be on it. Get well away, after having made sure that it can be found again, and leave completely alone overnight.

VEGETABLES AND SALAD. The seed grains of grasses make a good, sustaining food if boiled first but can be eaten raw, as can the pale tips of young grasses which are also nourishing. Stinging nettles must be boiled for about ten minutes before they become edible. Cooked this way they are quite pleasant tasting and very good food value. Many leaves can be eaten, so long as they are tender and are not acid or unpleasant to the taste. Watercress is to be found along most running fresh water and is good to eat. The taste is peppery, and the leaves should be thoroughly washed to remove parasites before eating. Bitterness, acid taste, a burning of the taste buds are all indications of poison. Any food found in the bush must be rejected out of hand if any unpleasant taste of this sort is found. Even when a food appears to be clean and safe it is best to start off by eating only a little and then waiting a while before eating more. Fungi, toadstools and mushrooms are best left alone. Only a few are safely edible and identification is always difficult if not impossible. Any red or reddish bush food — berry, nut, seed, leaf or whatever — should not be eaten no matter how appetising it looks. Red is the food colour that must always be avoided in the bush.

TRAPPING, except for the simple rabbit snare, can be a dangerous waste of time and effort, since elaborate traps usually require complicated arrangements of twigs, branches, string, nylon, vines, bark or other materials. Because of their complexity they are usually unreliable in operation. In any case, snares should only be improvised in an acute emergency. Their need is remote since the human body can go many days without food. Finding water is much more important since dehydration can lead to death in only a few days.

CHEWING GUM is found in the bush in the gum exuded from some wattles. It has only a little food value but the chewing reduces the feelings of hunger and thirst and prevents the jaws from atrophying.

Hunger is the best sauce for bush food.

Further Reading

Calvin Rustrum, *The Wilderness Route Finder*, Collier Books, 1967.
A.B. and J.W. Cribb, *Wild Food in Australia*, Collins, 1975.
L. Lannoy, *The Australian Bushcraft Handbook*, Horwitz Grahame, 1983.
Abbot and Mullins, *Bushcraft*, New Zealand Mountain Safety Council, 1984.
M. Dunlevy, *Stay Alive*, Australian Government Publishing Service, 1981.
Hobart Walking Club, *Safety in the Bush*, 1986.
St. John Ambulance Brigade: First Aid Manual, latest edition.

CHAPTER 13

ORIENTEERING

Oh! where is my wandering boy tonight?
The boy who is bravest of all.

ANON.

Excuses can no longer be made for map-and-compass illiteracy, not now that the art and science of land navigation has been modernised, simplified, and made interesting and enjoyable in a sport that everyone can practise.

ORIENTEERING, which is something like a car rally, but conducted on foot, began in Sweden in 1918 and has spread, slowly at first, much more quickly in recent years, to the rest of the world. It is exceptionally popular in Scandinavia and has become a part of Swedish national life. In 1963 a single orienteering contest in Sweden involved 182 000 people! People of both sexes and all ages can take part. The speed at which contestants move is their own choice and can vary from a pleasant saunter to very fast distance running. Orienteering is the best possible training and practice in map-and-compass skills. No one unskilled in navigation should venture into the bush without having first practised orienteering.

BOTTOMS UP! If, while enjoying a walk through the bush one day, you should come across a group of men, women and children kneeling on the ground with their bottoms sticking up in the air, while copying, with red pens, from apparently sacred documents scattered on the ground, you would not have stumbled across a new and esoteric nature religion but upon a group of orienteers about to begin an orienteering competition. Somewhere in the immediate vicinity of these totally absorbed and concentrated people, all pointedly ignoring your distracting presence, you would be sure to find a tent or two and a couple of large banners with the words 'Start' and 'Finish' written on them in large letters.

Usually there are two tents. One is the recorder's tent where the times taken by the contestants over the course are noted down, and the other is the registration tent where everything begins. Here the orienteer fills in a registration form, pays a small fee and opts for the course over which he wants to run, stride, jog, stroll or dawdle. At most Australian orienteering meetings there are four courses graded in length and difficulty from the easiest, novice or D course, to the longest and most difficult, A course. The D course may be 2 or 3 kilometres in length, or less, and the A course may be 12 kilometres or more. Amongst other things the form you fill in asks for your car or motorcycle registration number. This is a safety precaution so that when the day is over a check can be made on anyone who has gone missing or got lost. The registration form also makes the point that competitors *must* report to the 'Finish' tent whether they have completed the course or not. This is essential to avoid searches being mounted for people who have given up and gone home without telling anyone.

Having filled in a registration form and paid an entry fee, the orienteer is given a map, usually about 20 x 25 centimetres in size, of the immediate area where the courses to negotiate have been laid. The map may be printed in black-and-white or in up to four colours. To protect the map against rain, sweaty fingers, mud blotches, sweat dripping from the end of the nose, and other disfiguring stains most orienteers carry with them a quarto- or foolscap-sized, heavy duty, clear plastic pocket into which it can be slipped. Most stationers stock these pockets which are quite cheap and are useful for carrying and protecting maps of all sorts. For orienteering it is important to buy the heavy duty plastic sort as this provides a rigid enough surface on which to lay the compass without any support other than the hands. Despite its rigidity the pocket can be rolled and carried comfortably in only one hand.

Map and compass are absolutely essential; the plastic pocket can often be done without. Compasses may be hired cheaply at most orienteering meetings. Nothing more is needed for orienteering except comfortable old clothes and suitable shoes or boots; and the simple knowledge that follows.

After filling in the registration form the orienteer is also given a competitor's card divided up into numbered squares, a list of controls and a starting time. It is a good idea to ask for a starting time about twenty minutes to half an hour in the future. This gives reasonable time to study the map, enter on it any corrections that may be posted up near the registration tent, and build up a picture in the mind's eye of the sort of country that will be worked over. The sorts of things to look for are the general slope of the land, the most prominent features, the direction taken by streams and rivers, and the presence of potentially dangerous features such as cliffs or swamps.

At the 'Start' banner the seconds are ticked off by an official, and then the orienteer, map and compass in hand, moves off. The first stop is only a few metres, seldom more than a hundred, along the trail, usually clearly marked out, to the master maps area. Here are displayed, usually on the ground in protective polythene folders, several copies of the map for each of the courses. The courses are marked by red lines which join together numbered red circles where control points occur. The orienteer copies the course as accurately as possible onto a personal map and slips it into the plastic folder.

The orienteer stands up, puts the edge of the baseplate of the compass on the red line to the first control point to be reached that has been copied down from the master map. The compass housing is then rotated until the north pointing end of the needle is lined up with north on the housing and the direction of travel is read off from the big direction line arrow on the front of the compass baseplate. The direction of travel may now lie towards almost impenetrable bush a few metres away. After studying the map more closely the orienteer may find that it is possible to reach easily a nearby path that will lead quite near to the first control point, though the distance will be twice as far as through the dense bush. The situation is complicated by the fact that in order to reach the path the orienteer must push through a small area of dense bush anyway. Also, once in the vicinity of the control area it will be necessary to take a bearing, leave the path, and plunge into thick bush to find the red and white marker of the control.

The very first move confronts the orienteer with a fascinatingly complex problem about which a quick, logical decision must be made based on all the information that can be brought together from the map, the terrain, and the orienteer's own speed and physical capacities. The navigational sharpening-up process has already begun.

Finally, it is decided that the best solution is to take the path, which will be quicker than a bash through the bush. Arrived in the vicinity of the first control and still on the path, the orienteer finds that one set of problems has led to another. The control is deep in the bush and well away from the path which now, for the first time, stretches straight ahead. There are no features from which a bearing might be taken on the control. A quick look at the map provides a solution. The scale of the map indicates that the control is off to the right, at a right angle to the path, about 80 metres ahead. Knowing how many paces it takes to cover 100 metres the orienteer quickly works out how many paces will cover 80 metres and sets off. At the end of counting an abrupt turn to the right and into the bush is taken. About 40 metres in, and out of sight of the path, there is the red and white triangle that is the international orienteering marker for a control. But is this the right control? The orienteer consults the letters on the edge of the control sign. They coincide with the letters marked down for the first control point on the list accompanying the map. This is the correct control! The orienteer pulls down the punch hanging from the control and stamps the first square of the competition card, which is proof that the control has been visited. Immediately, the orienteer steps away to take a new bearing on the second control and a new set of navigational problems.

On the more difficult A and B courses there may be a dozen or more controls set out over 12 or more kilometres of varied terrain. The C and D courses are equally various, but easier and covering shorter distances. All courses are enjoyable activities as well as excellent training in safe, accurate navigation. Orienteering also teaches techniques that are useful anywhere in the outdoors and gives rigorous practice in applying them. 'Aiming off' and 'attack points' are only two applications of everyday orienteering techniques.

AIMING OFF is a technique used when the control ahead is on a more or less continuous feature such as a road, river, fence or footpath. The navigator will aim for the control but will more probably arrive on one side or the other of the control somewhere along the feature, and be unsure which way to move along the feature in order to reach the control. If the decision is wrong it will cost valuable time. In this case the orienteer 'aims off' deliberately to one side or other of the control. Then, on reaching the feature, will know definitely which way to move along it to reach the control. There is no indecision and time is saved.

Aiming-off. The objective (2) is on a stream junction and the ground between the starting point and the objective is fairly thick bush. The route taken is on easier ground to an area downstream of it. When the stream is reached, the navigator knows that he must move along it to the left (north) to reach his objective.

ATTACK POINTS are features in the immediate vicinity of the control. Instead of aiming for the control the orienteer aims at a suitable, easily found feature instead. On reaching it a meticulous bearing on the control is taken and followed carefully for the last 100 metres or so. In this way it is possible to pinpoint the control from close-to and arrive almost exactly on it. There are many other forms of orienteering but cross-country is the most useful and important.

STEP COUNTING. Many orienteers keep account of the distances they have covered by counting double paces. Every time the left

Attack points: When navigating over long distances it is often best to make for a distinct feature in the vicinity of the objectives and then navigate into the objective from this attack point. In A the attack point chosen is the corner of the dark area which is a field. The extra distance covered is negligible. In B the stream bend is the attack point. A longer distance but still reliable. In C the navigator aims directly for the objective knowing that if he misses it he can keep going to the road. From the road he can navigate accurately back towards the objective.

foot hits the ground it is counted as a pace and it is a simple matter to find one's own pace count over 50, 100, 200, 300 metres and just as easy to work from them over longer distances. Top orienteers count every pace of every course they cover. The double pace originated with the Roman legions. A thousand Roman (double) paces made up the *Mille Passus,* which later became abbreviated to 'a mile'.

It seems worth saying once more, and finally: orienteering is the best possible practical training in land navigation. In a few orienteering meetings more problems occur than in several years of ordinary outdoor navigation. Orienteering helps you to come to terms safely with the bush and the land. Navigational skill is only a small part of the deep respect we should have for the bush. Most Australians love the bush, but this is not enough. It has also to be feared and respected as sailors love and fear and respect the sea. Returning to the bush and nature, however temporarily, is a genuine pioneering effort demanding a competence in basic skills which comes only with practice and a deeply felt respect for the hazards and dangers of any pioneering endeavour.

The first essential basic skill is to be able to find your way about safely, so that you are wholly responsible for yourself and unlikely, barring unforeseeable accidents, to become the responsibility of search parties made up of people who can ill spare the time or energy to retrieve you from your foolishness or mistakes.

Knowing *where* you are is almost as important as knowing *who* you are, but very much less difficult.

In the outdoors always carry a map and compass.

Remember always, if you can't answer the question 'where am I?', you shouldn't be where you are.

Further Reading

Rand and Walker, *This is Orienteering*, Pelham, 1976.
A. C. Black, *Orienteering*, 1985.
J. Disley, *Tackle Orienteering*, Stanley Paul, 1984.
K. Haasma, *Orienteering*, Methuen, 1984.

CHAPTER 14

BIRDS

When birds do sing, hey ding a ding, ding . . .
WILLIAM SHAKESPEARE

When, in December 1696, Willem de Vlamingh, the Dutch
Commander of three ships sent to the coast of Western Australia to
seek survivors of a ship reported lost, arrived off Western Australia,
he saw a sight that must have filled him with astonishment and awe.
Anchored between an island and the coast, he ventured by rowing
boat up a large river nearby, where he and his men were the first
Europeans to see creatures that had hitherto belonged only in
classical Greek mythology: *black* swans. Several swans were caught
and two were taken back to Batavia in the Dutch East Indies.
Vlamingh named the river, 'Black Swan River' and it is known today
simply as the Swan River.

The explorers who later opened up the continent were also
interested in bird life, not for mythological associations but for
gastronomical ones. Ludwig Leichhardt was famous for the eagerness
with which he tried cooking and eating the local fauna. Fried emu
was a particular favourite of his. He and his companions sliced the
meat and dried it for preservation and portability, stewed the heads
and necks, and cooked even the feet and gizzards! 'Eating crow' is a
phrase we have come to understand as something like 'eating
humble pie' but probably neither phrase would have impressed
Leichhardt, who greatly enjoyed roast crow. In one journal
Leichhardt records a Christmas dinner for which one of the chief
dishes was stewed cockatoos. Perhaps absence of prejudice played a
part in his enjoyment as it did in the enjoyment of some new
Australians who in more recent years found that the kookaburra or
'laughing pigeon' was a bird very much to their taste.

Whether interest is mythological, gastronomical, or simply
ornithological, the Australian nest is well feathered with birds, from
the fourth largest of the world's predatory birds, the wedge-tailed
eagle, to the gimlet-eyed bulky strangeness of flightless birds like the
emu, and the emu-like cassowary, and the minute, thimble-sized
darting delicacy of the bush wrens.

THE CASSOWARY is a bird of the tropical lowland rainforest and is
so well camouflaged that it is difficult to see. Disconcertingly its
footfalls are heard first and then the quiet grumble-rumble of its
voice. The cassowary is best left alone. It manoeuvres well at speed
and attacks with powerful kicks, using the long, sharp nail of the
inside toes.

THE EMU is much more common in untouched quiet areas of the south and east than is generally realised. Emus live in pairs and flocks and are aggressive in defence of their young. They can run at speeds of up to 60 kilometres per hour. Their numbers are being steadily reduced by destruction of their habitat and the spread of fencing. At about 2 metres in height they are Australia's largest birds.

THE WEDGE-TAILED EAGLE, too, is more common than is realised, and is found over large areas of the continent. It has only begun to be valued as a pest-destroyer and carrion-cleaner in recent years, having been wrongly convicted of lamb killing on the basis of its fierce looks rather than its actual performance. Wedge-tailed eagles are able to make use of thermals (updraughts) of air to soar at great heights, for many hours. They hunt in pairs and are often intelligently co-operative.

Birds are fascinating creatures, whether like the sulphur-crested, or white cockatoos they are raucously announcing, and then maddeningly following, your progress across the countryside, or whether like the kookaburra they are choosing exactly the worst, sweat-trickling-into-the-eyes moment of finding oneself momentarily bushed to deride with great peals of harsh, ironic laughter. Birds are omnipresent in the bush and large numbers are active and visible during the daylight hours. They are fascinating in their behaviour and flight as well as often aesthetically deeply satisfying in their colouring and conformation. The coasts would seem very strange without the herring and Pacific gulls, and the cultivated land and the roadsides are hard to imagine without the presence and sharply musical fluting whistle of the magpie.

Flight, the fantasy of man until this last century, was achieved by much inferior creatures about 140 million years ago when two entirely separate animals launched themselves fairly successfully into the air. The pterodactyl was a gliding reptile and probably outnumbered the world's birds, but was destined to die out. The true ancestor of the birds was a semi-reptile, believed to be warm blooded, having teeth (modern birds do not have teeth), having a long jointed tail and skeletal similarities to the reptiles. This creature, a fossil of which was found in Bavaria in 1861, has been called *Archaeopteryx*. It had feathers and feet adapted to perching. As evolution progressed a few birds lost the capacity for flight but the others multiplied, spread, and adapted themselves to peculiarities of the environment from the polar areas to the tropics. Gliding, as distinct from flight using articulated wing movement, developed in a number of small mammals, some of which, like the sugar gliders, are common in Australia; and the flying fish in the warmer oceans.

The air is the kingdom of the birds, and flight is their single great superiority over nearly all the other creatures of the earth. Escape from the creatures of land and sea and rapid movement to

evade natural catastrophe, as well as the ability to swiftly seek out
suitable environments for survival, all depend on this single highly
evolved ability. Aerodynamically, the only thing birds lack is a
vertical fin for steering. This is accomplished by simultaneous
movements of wings, body, and tail. In flight the bird uses the
aerofoil shape of its wing in a downward and forward motion so that
uplift ensues and the bird moves forward and upward. Timing and
wing flexing greatly reduce drag. Flapping flight makes great
demands on a bird's energy. Few birds flap their wings continuously.
Most glide whenever possible, taking advantage of updrafts of warm
air called thermals to help them soar upwards with a minimum
expenditure of effort. Glider pilots use exactly the same tactics with
their machines. The best gliding birds, the albatrosses, like the best
gliding machines, have high aspect-ratio wings — long, narrow and
tapering.

Birds serve man well. The main food groups for birds are insects
(the largest), nuts, seeds, berries and fruit, grass and herbage, flesh,
and scavenged material. Birds consume and destroy enormous
numbers of harmful insects, pests like rabbits, mice, and rats in their
search for food. They keep the beaches and roads clean, and eat huge
amounts of the seed of weeds. Birds play an important, and seldom
appreciated, part in the natural ecology of the land. It is only when a
species comes near to disappearing point that this is understood.
Equally, interference with the ecology by the deliberate introduction
of pests like the sparrow and the starling, as has happened in
Australia, can dangerously disrupt the existing ecological pattern.

BIRD WATCHING can be anything from a casual entertainment to
an intensely serious study carried on with specialised cameras
and (heavy) telephoto lenses, binoculars, sketch books, hides, and a
variety of handy paraphernalia. For the bush venturer, birds are a
delight to the eye and ear, and identification is what he is mostly
concerned about. This is more difficult than it seems because, despite
the excellent bird books about, there are a number of problems. Bird
plumage varies in both female and male birds in the young, and in
the course of the year. Some plumages change distinctively in the
year leading up to maturity, some go through distinct phases. Shades
of plumage colour vary enormously. What is described as yellow in a
book may be anything from a watery whitish yellow to a blindingly
bright lemon-yellow. Birds are intensely mobile and are seen much
more in flight than stationary. Very few have identification marks
that are unambiguous at a distance. In flight, birds present a
silhouette to an observer more than a picture.

Despite all this, accurate identification takes place all the time,
largely on the basis that if you want to learn to identify birds you
have to go and identify them. This means keen observation all the
time and the building up in the memory of a composite of

information so that silhouette, or size, or distinctive marking, or song, or gait, or peculiarity of flight make identification accurate, effortless and instant.

MIGRATION is the regular seasonal movement of birds from one locality to another, usually from cold to warmth and back again to the cooler area which is where breeding takes place. Migrant birds come to Australia from as close as New Guinea (shining starling, nutmeg pigeons), to as far away as the Arctic (golden and grey plover, curlew, whimbrel, sandpiper). Others come across the Tasman from New Zealand (double-banded dotterel) and across Bass Strait from Tasmania (pallid cuckoo, eastern silvereye, tree martin). From the intensive cold of the Antarctic to the mildness of the Victorian and Tasmanian winters come albatrosses, petrels, skuas, rockhopper penguins and others. Within Australia migration occurs from north to south and from east to west, over great distances or over only small distances. A few birds are not migratory but nomadic (budgerigars, wood-swallows) and do not return to the same place to breed. There are innumerable birds which remain in the same area permanently.

TERRITORY is of fundamental importance to bird life. Nature deals ruthlessly with the young of birds at all stages of their lives but especially in the earliest days of eggs and chicks. It then ensures that the mature birds are evenly dispersed across their winter territory so that they seldom exceed the food supply that is available, and competition is reduced. Mature birds guard their territories zealously. It is often not realised that bird territories are made up of a complex structure within which survival can be assured for most of the time. Some types of bird territory are: individual distance territories, breeding territories, family, and flock territories. The distance between individual birds, between pairs, and between groups is strictly structured and rigidly maintained. Winter territories are usually the most important because they involve food supply. Birds are indefatigable in defending winter food territories,so much so that some birds have been known to keep so busy in repelling invaders that they have starved to death amid an over-abundance of food.

There are well over seven hundred birds native to Australia, without mentioning the many migratory birds that visit our shores, or the many birds which do not breed here. Many Australian birds, amongst them the emu and the cassowary, are now believed to have been inhabitants of the continent since Australia separated and moved away many millions of years ago from the single huge landmass existing at the time. Large families of birds, including most of our wrens, parrots, cockatoos, and honeyeaters, are unique to Australia. In a single chapter it would not be possible to mention all, or even most Australian birds. Instead a selection has to be made, and that of birds most commonly seen in the most common natural habitats outside the cities of the east and south.

HONEYEATERS are a large, widespread Australian family of birds whose livelihood is assured by the abundance of flowering trees and shrubs, though most of them eat insects as well as nectar. Some plants depend on the long, curved-bill honeyeaters for their pollination. They are small birds usually around 15 centimetres in length. One of the best known of the Australian birds, the bell-miner or bellbird is a honeyeater. Bellbirds live in flocks in both underbrush and tall timber on ridges and beside creeks. They live in part on nectar from flowers and in part on leaf-insects. The New Holland honeyeater is the bird that betrays the intruder in the bush, making noisy calls of alarm and following up with fussy, nagging chattering. This long-billed bird with a black-and-white striped breast, and golden patches on its folded wings is also known as the white-eyed, and yellow-winged honeyeater. In eastern Australia it lives mostly on the east side of the Great Divide from southern Queensland down to the south-west of Victoria, and westward of the Nullarbor plain. The bird of the south-west of Western Australia is quite different from the bird of the east.

The Lewin honeyeater is a bird of the rainforest and is only incidentally a nectar eater. Most of its diet is made up of fruit and insects. It is found from Cape York right down to the south of the east coast. It is olive-green in colour and merges well with the leafy background of the rainforest. There are yellow streaks along the edges of the beak and bolder yellow ear markings.

The yellow-throated miner which is also known as the white-rumped miner is a medium-sized honeyeater widely spread through the drier inland areas. These are noisy, obvious and busy birds which associate in small flocks and family groups. They respond noisily to any threat and are brightly alert to everything that is happening around them. When one bird gives a cry of warning the others join in a harsh chorus. Food is mostly found at ground level and nectar plays a subsidiary part in their diet.

WRENS. Fairy wrens, with their narrow vertically outstanding tails, tiny bodies and colourful plumage, are some of the most delightful and charming of Australian birds. They are pugnacious and have been often seen vigorously attacking their own reflections in glass, and are amazingly industrious. A study of a pair of blue wrens showed that in a single day they carried 2800 insects to their nest as food for their young! They are cheeky, active and irresponsible birds.

The superb blue wren is the best known and most widely distributed of the fairy wrens. It has a surprisingly loud, lively, chattering voice, flies in a straight line in short spells, and feeds mostly on insects found at ground level. These birds can be found easily in thickets and along water courses, as well as in gardens and orchards. They usually stay in one locality all their lives unless conditions make another area more suitable.

The variegated wren, which is also known as the purple-backed wren, carries a variety of plumage colours. Its head is purplish-blue, its shoulder chestnut, tail blue, chest and neck black, abdomen white. It is one of the smallest of the fairy wrens and has a softer voice than the blue wren. It is common throughout the whole of eastern and south central Australia.

PARROTS form one of the largest groups of birds in Australia. There are about sixty kinds, which include cockatoos, cockatiels, corellas, honey parrots, fig parrots, rosellas, grass parrots or budgerigars, and ground parrots. From the earliest discovery of Australia parrots seem to have made a deep impression on explorers and visitors. Captain Cook noted in his Journal 'Cockatoos of two sorts White and Brown, very beautiful birds of the Parrot kind, such as Lorryquets'. As early after settlement as 1839 John Gould, one of the greatest pioneer naturalists, was writing: 'No group of birds gives Australia so foreign and tropical an air as the numerous species of this great family, each and all of which are very abundant . . . Nay, the very towns, particularly Hobart Town and Adelaide, are constantly visited by flights of these beautiful birds, which traverse the streets with arrow-like swiftness'.

Parrots and cockatoos are easily recognised by their strong, stubby, hooked beaks and by the two toes in front and two toes behind their feet, as well as by their vivid colouring. Some parrots have brush-tipped tongues which allow them to extract nectar from blossoms. The dominating colour of most of the honey parrots is a relatively subdued green. Cockatoos have particularly powerful beaks and some can crack nuts which almost defeat the leverage of metal nutcrackers. Cockatoos often use their beaks to pull off bark and break open wood to get at grubs. White cockatoos are usually left-footed!

There are eight species of rosellas which are amongst the most brilliantly coloured of the parrot group, with dominant colours of scarlet, yellow, and bright green, which make their identification, as they erupt swiftly from concealing foliage, uttering their shrill shrieks of alarm, a simple matter. When feeding they make only a soft scratching-sounding sort of chatter.

Grass parrots or budgerigars are limited to the warm-to-hot interior and usually to the west of the Great Dividing Range. These small birds have exceptional powers of recovery. In 1931 a heatwave hit Central Australia and budgerigars perished in millions. One report spoke of 'five tons' of birds being drowned in a dam into which they had plunged to escape the searing heat. Other reports spoke of '30 000 dead birds' and '60 000 dead birds'. A year later there seemed to be just as many budgerigars about as ever.

There are only two corellas, the little corella and the long-billed corella. The little corella is widespread across much of the interior to the west of the Great Divide, and the long-billed corella is limited to two small areas in south-east and south-west Australia. Both birds are predominantly white in colour and undistinguished in appearance, but both make good pets and are welcomed by visitors to the inland because they are a reliable guide to water.

BUTCHER-BIRDS, MAGPIES AND CURRAWONGS all belong to the same family. The butcher-bird gets its name from the butcher's shop it sets up, by wedging its prey in a tree-fork or branch, or spitting it on a large thorn, to be butchered or for safe keeping. The Australian magpies, which are strongly but superficially similar in appearance to the European magpies, are really a variety of butcher-bird. The European magpies are members of the crow family. Butcher-birds and magpies are bold and aggressive. They will attack each other, dogs, cats, goannas, and anything that moves, including human beings. Currawongs, or bell-magpies, have been known to attack and injure human beings, in defence of their nests and young. Most birds of this family have remarkably pure, flute-like whistling voices. The two most widespread of the butcher-birds are the pied and the grey butcher-birds. Magpies, either black-backed, or white-backed, are to be seen over most of the continent. The pied currawong is distributed widely along the east coat from Cape York to Victoria.

KINGFISHERS, apart from the kookaburras, fall into two groups: birds of the water or birds of the wood or forest. There are eight kinds of kingfisher in Australia, the most common of which is the sacred kingfisher. All the kingfishers are brilliantly, iridescently coloured, and have long beaks which they often use to spear their prey. Kingfishers swoop on their prey from either a perching or a hovering position. Like kookaburras, the other kingfishers nest in termite mounds which they drill into and enlarge and excavate with their beaks, labouring in pairs.

The sacred kingfisher is only about 20 centimetres long and is found where there is open wooded country and water, and in mangroves along the coasts. The colouring is predominantly brilliant blue and white. It lives off worms, grasshoppers, lizards, beetles, and dives for fish and crabs. Captured prey is carried back to the perch where it is beaten to death before being eaten. Pellets of indigestible matter — shell, bone, insect wings — are regurgitated. The sacred kingfisher is a migrant bird in the south, which it leaves each winter to travel north into tropical Australia and further afield to the Indonesian islands. Some of the birds migrate to New Zealand.

The azure kingfisher is always found along streams and in the vicinity of water in northern and eastern Australia and Tasmania. Its colouring is a sparkling combination of bright ultra-marine blue, rusty red, and white. Its flight is swift, close to the water, frequently altering course with darting dexterity. The perch is usually close to the water and concealed in the vegetation. When perching it makes frequent small movements of its head and feet which make it easier to spot. This kingfisher dives for food which is made up of small fish, shellfish, frogs and water creatures.

The red-backed kingfisher is widespread over most of the continent. It has a bluish green body with white underparts and a distinctive reddish rump. It is a nomad and migrates to various inland parts of the continent in winter; the only kingfisher found in the arid centre, it can be seen far from water. The dead limbs of trees are preferred as perches and food may be insects, lizards up to 20 centimetres long, scorpions, centipedes, and insect larvae. Nests are made in hollow trees or in termite mounds.

KOOKABURRA is the Aboriginal name for the two largest of the kingfisher family, the laughing kookaburra and the blue-winged kookaburra. The laughing kookaburra is the bigger by about 1.5 centimetres and grows to about 46 centimetres in length. It has a large white spot on each wing, clearly to be seen during flight, and a dark brown strip the width of the eyes, extending backwards behind them. The eyes are not especially sharp but they are acutely sensitive to movement which enables the hunting bird to locate the small movements of prey, a small snake, lizard or insect, from many metres away. As its beak and claws are not made for killing prey, a kookaburra's neck is immensely strongly muscled. This bird beats its prey to death, usually on the branch on which it has been perching, but occasionally on the ground.

The laughing kookaburra is widespread throughout eastern Australia and in the south-west corner of Western Australia. The mad raucous laughter from which it takes its name is most commonly heard around dawn and dusk, but is also uttered at other times as a danger warning. Kookaburras adapt to a variety of territories and in the vicinity of rubbish dumps they become scavengers. Like owls, kookaburras regurgitate the unwanted bones, fur and feathers of their prey in the form of pellets which gradually pile up beneath the trees they roost in at night. There have been reports of kookaburras killing snakes up to 1 metre in length, and there has been one instance of a pair of kookaburras attacking and killing a carpet snake almost 3 metres long. They take freshwater crayfish and yabbies by diving into the water, but they are clumsy fliers, sometimes drowning in water tanks by skimming low over the surface to drink, and being unable to rise quickly enough to avoid colliding with the far side, falling in and drowning. Kookaburras nest in hollows in trees or high up in holes in large termite mounds.

The blue-winged kookaburra is a very close relative of the laughing kookaburra and is similar in size and appearance, though not in colour. The blue-winged kookaburra is confined to northern areas of the continent and is easily recognisable by its call, which is a dog-like yelping bark. Its Aboriginal name is *Pooka* which means stinking.

WARBLERS. Small birds that are brownish in colour make up a good deal of the bird population, and are hard to identify. This relative drabness of plumage is true of many of the birds which make up the group of birds known as warblers. Most feed amongst the foliage of trees, in undergrowth, or on the ground. Insects make up the bulk of their food. Warblers can be found all over the continent and in every sort of habitat, under even the most extreme environmental conditions of desert, coast, rainforest, or alpine country, as well as the easier conditions of sclerophyll forest, farmland, and city parks and gardens.

The rufous songlark, a quite small bird of about 18 centimetres in length, is one of the warbler group. Its chest and underparts are off-white, tail dark brown, back and sides grey-brown. It has a white eyebrow. Its habitat is most commonly dry open woodland throughout Australia but not Tasmania. The rufous songlark migrates from southern areas in late March and does not return until the following August. In particularly mild winters many birds stay in the south. Its food is larvae, insects and some seed. It is one of the most common of Australian birds.

The brown songlark grows to about 25 centimetres long and is generally a streaked dark brown in colour. His large size, and almost vertically cocked-up tail when perching, make identification quite easy. The brown songlark is a bird of the cultivated paddocks and open plains. It perches in bushes and on overhead wires and fences and posts. Its food consists of beetles, caterpillars and seed. Like the rufous songlark it migrates north in winter unless conditions are exceptionally mild.

The yellow-tailed thornbill occupies parks and gardens, orchards, paddocks, grass plains and open forests over much of the south and east. It is a small bird about 11 to 12 centimetres long, of mixed colouring: forehead speckled white on black, chest whitish, back pale brown, rump bright yellow. The bright yellow rump shows up particularly well in flight. It is a bird of the ground and quite common. Gardeners welcome this little bird which eats insects, larvae and various garden pests, though it is an easy prey for domestic cats.

The weebill, at 8.5 centimetres in length, is one of the smallest of the warblers as well as one of the most widespread over the continent. It prefers open forest, mallee, and forest where there is an abundance of saplings and small trees. The back is greenish yellow, wings brown, underparts pale yellow, tail brown. The weebill is a brisk, active little bird, ceaselessly moving through foliage after food which is mostly insect life. These birds are often difficult to watch because of their small size and the speed at which they fly.

The white-throated warbler occurs from Cape York, down the east side of the Great Divide, to southern Victoria but not Tasmania. Its throat is white, upper parts ash grey, lower parts bright yellow. The habitat preferred is more often open timbered country in mountain areas as well as lower down. This bird is a migrant to the south-east from September to March or April and feeds exclusively on small insects and larvae from foliage. The song is loud and pleasant and has led to the bird being called the bush-canary.

FLOWER PECKERS are brightly coloured small birds with stout beaks, short tails and stumpy bodies. They get their group name from the fact that they eat fruit, nectar, mistletoe berries and insects. There are two sorts of flower peckers. Pardalotes are exclusive to Australia, and the mistletoe-birds are a group of their own.

The mistletoe-bird is a very small bird, only about 8.5 centimetres long. It is scarlet-breasted and scarlet under the tail with glossy blue-black back and whitish underparts with a central black stripe. This bird is found wherever there is mistletoe growing and so covers the whole of the continent except Tasmania. It feeds on other berries as well, especially Japanese pepper, and a few insects. Its flight is speedy and irregular in pattern.

The striated pardalote has a striped head and a broad white stripe in the wing. Its upper parts are grey and buff, wings black with a red spot at the base of the white stripe. The eyebrows are orange-yellow. This bird is found over most of the middle and southern areas of Australia mostly in drier country amongst trees, where it nests in hollow trunks, but also sometimes digs tunnels in which it nests. It feeds mainly on insects amongst the foliage of eucalypts. One spotted pardalote is limited to King Island and Tasmania.

FLYCATCHERS, do, in many cases, take insects on the wing, but most of these birds take their food from the ground and from foliage, branches and tree trunks. Flycatchers move quietly and unobtrusively at all levels from the crowns of trees to the ground. The two main groups are the Australian robin and the fantail. The Australian robins have no connection with the European robins. One of the best known of the flycatchers is the willie wagtail, a fan-tailed, black-backed and white-breasted bird with black head (and throat) and white eyebrows. Its length overall is usually about 20 centimetres and it is common throughout Australia and neighbouring islands. It is common in parks and gardens, almost anywhere in forests, along the watercourses of the dry inland, and on cultivated land. The fantail is in constant swaying movement.

The restless flycatcher is white underneath, including the throat, and glossy black on head and back. It is like the willie wagtail in appearance and size — 20 centimetres — but does not have the prominent white eyebrow markings. This bird prefers lightly wooded areas, pastureland, and orchards, as well as light scrub. It is found throughout the whole of eastern and northern Australia and in the south-west of Western Australia. Feeding on insects takes place on the wing while it hovers quite close to the ground. It also takes moths, caterpillars and butterflies. This is a restless bird constantly on the move. It is also known as the scissors grinder and dishwasher.

The scarlet robin is one of the most attractive small birds of south-eastern Australia. It is about 13 centimetres long, has a jet black head, back upperparts, and throat, a bright scarlet breast, and a very noticeable large white spot on the forehead. This robin keeps largely to the dry sclerophyll forests and open woodland of the coastal regions, but is also quite commonly seen in orchards, gardens, parks and on cultivated land. It lives on beetles, moths, larvae and insects taken usually from ground or foliage.

The flame robin, which is also widely found in south-eastern Australia, is sometimes confused with the scarlet robin which it resembles in size and superficially in colouring. Nevertheless there are distinct differences. The flame robin has a distinctly orangey flame-coloured breast and throat and has a sooty-grey back. The white spot on the forehead is quite small. This bird is often migratory, moving backwards and forwards from Tasmania, to and from high country, where it can be found in flocks on the high plains above 1200 metres, and between heavy and light timber. Food is mainly insects, larvae and beetles taken on the ground. Stubbles and open grassland are often frequented.

The hooded robin is a smallish, 16 centimetres bird with contrasting black and white plumage which makes it look like a miniature magpie. In flight it shows a distinct white wing bar. It is distributed widely throughout most of temperate Australia and in parts of the Northern Territory. It prefers drier inland savannah country, though it is found in a variety of other habitats. Food is entirely insects gathered from the ground in a nomadic existence. Areas of dead or fallen timber provide much of its food.

The eastern whipbird is also a flycatcher. It is about 27 centimetres in length. The back, wings and tail are a beautiful deep olive-green. The breast is black with a white spot. The head, with its distinctive crest is black, but the cheeks and the sides of the throat are white. This whipbird occurs through eastern Australia from north Queensland south to Victoria and is found in rainforest, high rainfall bush areas on the ranges, thick scrub and profuse luxuriant vegetation. Foraging for insect food involves the brisk turning over of ground debris and leaves. This is a shy bird. Pursuit is useless, but

BIRD GROUP	SIZE		HABITAT
Swifts Swallows	13-20 cm		Open spaces
Magpie-larks	25-28 cm		Streets & parklands
Flycatchers Fantails Robins	11-20 cm		Grasslands & woodlands
Whistlers Thrushes	18-28 cm		Gardens & woodlands
Thornbills scrubwrens	10-13 cm		Tree-tops & undergrowth
Fairy-wrens	10-13 cm		Heath undergrowth
Pardalotes	9-11 cm		Tree-tops
Honeyeaters	14-37 cm		Heath, scrub tree-tops
Larks	14-28 cm		Grasslands
Finches	11-17 cm		Grasslands heaths
Magpies Currawongs Ravens	28-48 cm		Grasslands parklands forest

BIRD GROUP	SIZE		HABITAT
Terns Gulls Plovers	23-66cm		Beach Paddocks Swamps
Cormorants	58-81cm		Waterways generally
Waterhens Rails	18-56cm		Swamps River banks
Grebes	25-51cm		Lakes Dams
Herons Ibises Spoonbills Egrets	28-91cm		Waterways
Ducks Geese Swans	41-102cm		Waterways
Hawks Falcons	30-97cm		Open spaces
Parrots Cockatoos	17-66cm		Open spaces woodlands
Pigeons Doves	30-36cm		Roadsides gardens
Kingfishers Kookaburras	20-43cm		Open forest streams
Cuckoos	15-30cm		Forests gardens

'decoy duck' tactics sometimes work. Sitting down quietly will draw the bird's curiosity and it may then come out to make an inspection and allow itself to be seen. The call of this bird is the well-known swift whistle rising to a sharp and unmistakeable whip crack. Once heard it is nearly always associated thereafter with the damp air of the rainforest, fern gully or humid bush.

BIRDS OF PREY are not always the inhabitants of wide, bare, open country, as is still commonly believed. They occur as well in rainforests, savannah woodland, and eucalypt forests. Hawk is the general name for predators called kestrels, falcons, harriers, kites, buzzards and eagles, all of which have markedly hooked beaks and large, strong, sharp talons, and which range from medium-small to very large in size. Depending on size, hawks feed on anything from large insects like grasshoppers to small reptiles and birds, and rodents and small mammals. The wedge-tailed eagle preys on rabbits, but is also a carrion eater as is the whistling kite. The peregrine falcon, which measures about 46 centimetres in length often seeks its prey amongst parrots in open woodlands. When the flock rises in fear the peregrine singles out a straggler bird and swoops at great speed to the attack. The prey is usually dead before it reaches the ground. Woodlands usually support more predators than forests.

Although it is not much smaller — 38 centimetres — in contrast to the peregrine the crested hawk lives almost entirely on insects. Grasshoppers, cicadas and beetles are only very occasionally varied with a small lizard or mouse. The brown goshawk is another forest inhabitant but is better equipped in claws and beak as well as being considerably larger at 53 centimetres. This goshawk flies extremely fast in pursuit of prey which is made up mainly of small birds and mammals. The strength and ferocity of this bird can be judged from the fact that it also takes rabbits, and native cats!

Hawks often hunt in pairs, combining to flush out their prey, only one seizing it. Most hawk nests are made high in trees out of sticks and leaves, usually in the shape of a flat platform or raft.

THE NIGHT HUNTERS of Australia are mostly owls. These birds have large, strong beaks and talons but also have very large eyes which give them exceptionally sensitive night vision. Like the hawks, owls are outstandingly beautiful birds whose compelling eyes, usually set in the front of the head, give them a rare, imposing dignity. Owls work in complete silence, even when the larger varieties take small mammals like ring-tailed possums. The talons penetrate deeply and quickly, and death is usually instantaneous. Owls glide silently to the attack, hardly seeming to disturb the air. They come and go as swiftly and silently as puffs of smoke. Owls have to be seen in action before they begin to exert their full fascination.

The powerful owl is the largest of the Australian owls at 66 centimetres. It is a handsome brown bird barred with white on the upperparts, with breast and underparts a buff-white barred with brown. It is large enough to fly easily, carrying possums and possum-gliders, and has distinctly orange-coloured eyes. Its habitat is the thickly forested ranges from Victoria to southern Queensland where it lives on a varied diet which includes flying foxes in the northern area, and possums, possum-gliders, kookaburras, magpies, domestic kittens (gone bush), rats and large beetles elsewhere. This owl is quite rare in Victoria where the birds are said to grow to a larger size than further north. In all areas the pairs of birds are most usually found in the vicinity of the overgrown gullies of timbered hilly country. Having fed, birds will sometimes roost for hours with the carcass of their prey dangling from their talons. Nests are usually in holes high in old part-hollowed trees on steep, heavily wooded slopes.

The barking owl is a medium sized, dark brown bird about 40 centimetres long. It is of special interest to people camping in the bush because of its occasional habit of producing very loud screams and moans in the middle of the night — usually in late autumn. Aside from these rare macabre cries, the barking owl has a gruff 'wook-wook' call that is almost indistinguishable from a dog's bark. The cry of the female is much higher and can be heard over much greater distances. These owls are found spread sparsely all over Australia except for the centre and part of the west. They enjoy a varied diet of possums, gliders, magpies, rabbits, young hares, rats, mice and small and medium sized birds. They are only partly night hunters and can find and strike down their prey in bright sunshine. Their intelligence makes them amenable to domestication, when they are capable of showing a great deal of affection. David Fleay, Australian authority on owls, has suggested that the barking owl could be trained as a hunting falcon for use both night and day.

Boobook owls are the most common of all the Australian owls, living in tropical jungle, eucalypt forests, savannah woodlands and arid desert. They are medium sized birds about 40 centimetres in length, dark brown in colour, streaked and spotted with white, with grey-green eyes. Their name originates from the Aboriginal, and describes their call, which is also described as a sing-song, slow 'more-pork'. The Tasmanian bookbook is smaller than his mainland relative, strongly spotted on its underparts, and is unique in that it is sometimes migratory, some birds crossing Bass Strait to the mainland in winter.

The boobook frequents city and suburban parks and gardens and plays some part in keeping down sparrows, as well as eating moths and beetles. Beyond the cities the boobook feeds on small rodents and reptiles, large insects and small marsupials. Another nocturnal bird, not an owl, the tawny frogmouth has acquired the name of the morepork bird because the cry of the boobook has been wrongly attributed to it. It is definitely the boobook owl which makes the 'more-pork' call.

THE LYREBIRD is the most picturesque of Australian birds, as well as the one that has received the most publicity. It was first sighted in 1798 when it was described as 'a bird of the pheasant species'. In the actual report of the shooting of the male 'pheasant', a wombat and a koala are also referred to for the first time! Governor Hunter sent an egg from the newly discovered bird to Sir Joseph Banks in London. The safe arrival of the egg was never recorded. Debate about the lyrebird went on for many decades. It was related to poultry, it was given 'from twelve to sixteen' eggs, which were white with blue spots, and these were laid on 'dried leaves or grass' within a hollow log on the ground. The truth is that the lyrebird builds a large nest of sticks, dry fern, and mosses, with a side entrance at anything up to 20 metres from the ground. It lays one egg only, brownish grey and streaked and spotted with dark brown and dark grey.

The two things that distinguish this large, dull-brown bodied bird of 102 centimetres in length, are its remarkable ability as a mimic of other bush birds and bush sounds and its lyre-shaped tail (on the male) which is displayed brilliantly to the female. Its habitat is rugged, heavily wooded country east of the Great Divide, and it is distributed from southern Queensland southward to the vicinity of Melbourne. The lyrebird lives on worms, insects, and small shelled creatures which it finds by picking at, and sifting through, leaves, debris and decaying logs. It is really a ground bird with only minimal flight which occurs in short bursts.

Further Reading

G. Pizzey, *A Field Guide to the Birds of Australia,* Collins, 1980.
Birds of Australia, Curry O'Neil, 1983.
J. Bransbury, *Where to Find Birds in Australia,* Hutchinson, 1987.
N. Cayley, *What Bird is That?* Angus and Robertson, 1984.
S. McCoy, *Australian Birds,* Collins, 1982.
Royal Ornithologists Union, *Birds of Eucalypt Forests and Woodland,* 1985.
Birds of the Australian Bush, Bay Books, 1981.
Blakers, Davies and Reilly, *The Atlas of Australian Birds,* Melbourne University Press, 1984.

DANGERS

*Man is born unto trouble, as the sparks fly
upward.*

JOB 5 : 7.

THE DANGERS OF THE BUSH are relatively few and are generally
much overrated. If anything they lie more in the people who go
heedlessly into the bush than in the bush itself. Bush creatures are
seldom aggressive but there are a few who respond to attack, or
apparent attack, with deadly weapons. Blundering carelessly along
isn't good enough; commonsense care should always be taken.
Almost all bush creatures will move quickly away from human
beings if they get the chance. It is only when they are surprised,
often when asleep, that they defend themselves by attacking. Good
bushmen are always acutely aware of the bush around and ahead of
them. They wear appropriate clothing and treat their surroundings
with respect. They never reach into rabbit holes or into hollow tree
trunks, but use a long stick and take the greatest care, just as they
invariably look carefully first at the trunk or branch they are going to
catch hold of in case it may be the resting place of a potentially
lethal snake.

THE RISK OF ATTACK remains minimal. The open bitumen roads
are far more dangerous than all the bush, mountains and deserts
put together. Thousands die annually on the roads of Australia, tens
of thousands are injured and crippled. Deaths from snake or spider
bites seldom reach more than a few each year, and those occur
usually in suburban backyards or on rural properties.

SNAKES are commonplace over most of Australia and in the
vicinity of great cities just as much as in desert or dense bush. Of
the varieties of snake in Australia only about sixty are venomous.
Snakes are shy, quiet creatures who want only to be left in peace and
so they usually live their lives as carefully concealed from human
beings as possible. Their weapons are a pair of curved fangs which
are grooved to allow poisonous venom to pass along them from the
salivary glands and into the tissues of the victim. The venom of these
snakes can have many effects, the most common of which is paralysis of the muscles, which in its final stages leads to asphyxiation
and death. It is because of the fast action of this neurotoxin that the
use of a tourniquet is the urgent and fundamental first step of any
treatment. Suitable clothing is the best preventive measure against
snakebite. When the fangs have to pass through clothing venom
spreads out into the clothing instead of into the flesh and the bite is

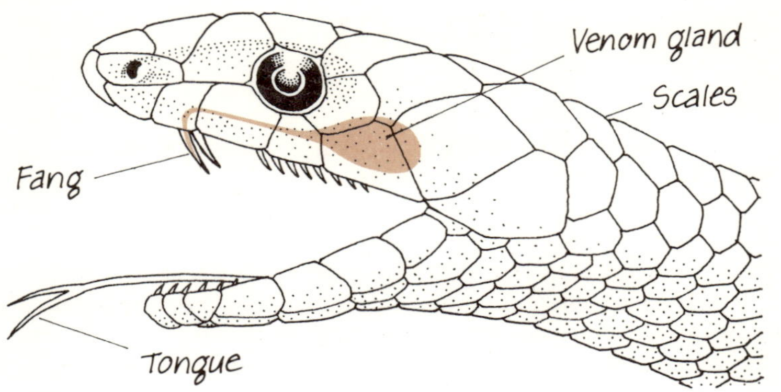

immediately made less serious. The more flesh that is bared, the greater the danger from any snakebite. Shorts and thongs or sandals are *not* suitable wear for the bush at any time.

The tiger snake is the most common of the deadly snakes. It is widespread in the southern half of the country but is not found much north of Brisbane. Tiger snakes vary in colour from grey through brown to olive with yellow banding. Some varieties of tiger snake are jet black. This species grows on average to about 1.5 metres in length. The fangs are long and the quantity of venom that can be injected is large. This is the most dangerous of the venomous snakes. It is found in numbers along river flats.

The copperhead inhabits the southern coastal ranges of eastern Australia. This is a thick snake, varying in colour from palest copper to black. The venom is as deadly as that of the tiger snake. Copperheads are often found in thick grass and tussocks along creeks and close to swamps. In warm weather they move mostly at night. They feed on lizards, frogs and other small creatures and are cannibalistic. They grow to 1 to 1.3 metres.

The death adder is common throughout the continent apart from Victoria and Tasmania. It is an especially dangerous snake because of its habit of attracting its prey by movements of the tip of its tail. It usually does this while lying concealed beside a track where it is easily trodden on or kicked and bites very swiftly in retaliation. Unlike most of the other snakes it does not move out of the way when someone approaches, making its danger to man proportionately greater. The colour is grey to reddish and it grows up to 1 metre in length. The fangs are long and the venom abundant and strong. Its movement is mostly at night.

The taipan is confined to the north of Australia, roughly from the Queensland border. It is the largest of the venomous snakes, growing to 2.6 metres. Even larger specimens have been recorded. Most taipans are dark brown in colour. It is one of the world's deadliest snakes.

The brown snakes are found all over the continent and include the collared brown snake, and the spotted brown snake or dugite. These snakes grow up to 2 metres long and move very quickly. They come in all shades of brown and may be blotched or spotted with other colours. These are very dangerous and will strike frequently if threatened.

The black snake has a distinctive red belly and is often seen swimming in rivers. It is a handsome creature which grows to about 2 metres long and is fairly common down the length of the east coast but not into Tasmania. This is an unusual venomous snake because when cornered or in difficulties it hisses loudly and sways and darts its head without seriously trying to bite. If it can get away it does so at high speed. The venom is not as toxic as that of other snakes but a bite is still a serious matter which should be treated just as seriously as a bite from any other venomous snake.

These are a few of the most common venomous snakes of Australia. There are many others. *Always treat snakebites as venomous and extremely dangerous,* until expert medical opinion says otherwise.

AVOIDING SNAKEBITE demands vigilance, suitable clothing — leather boots and gaiters will give excellent protection — and extra care, or the avoidance of places where snakes are likely to be common, that is around streams, creeks, rivers and swamps. Snakes, with the exception of the death adder, usually move away quickly as people approach, so stealthy movement, which gives no warning, is not the best approach to make. Snakes are usually out in warm, sunny weather.

The bite itself is a negligible thing almost without pain. The snake's fangs penetrate the skin like fine needles and leave little or no swelling. The distance between the punctures gives a rough idea of the size of the snake. One centimetre (just under half an inch) indicates a small to moderate sized snake. A very much larger and therefore more dangerous snake is suggested by a distance of 2.5 centimetres. Occasionally there are scratch marks on the skin from the snake's teeth, or reserve fangs, or a number of punctures if the snake has bitten several times.

Symptoms may appear anything from fifteen minutes to several hours after the bite. These show themselves in the muscles of the tongue, lip, jaw and eyelids which go into an unresponsive, flabby paralysis which gradually spreads to the chest and lungs, seriously affecting breathing, and leading if untreated to death from asphyxia. An early symptom is swelling of the lymph glands. Depending on the kind of snake, other symptoms may also show themselves: nausea, headache, drowsiness, sweating, diarrhoea, and most clearly, the passing of red or very dark urine. If there is marked swelling round the bites and intense pain is felt, it is almost certain that the bite is from some creature other than a snake.

In the event of snakebite, *don't* attempt to cut or wash the bitten area. Bandage firmly as much of the bitten area as possible (usually a lower limb or arm) with a broad bandage or strips torn from clothing. Immobilise the area with a splint or sling until medical help is at hand. Do not give the victim alcoholic drinks of any sort. Tea or coffee may be given. Keep the victim as still as possible. The broad bandage may safely be left on for several hours. Do not attempt to remove the bandage unless on expert medical advice. *Any* movement is likely to spread snake venom further through the system with possibly fatal results.

Psychological support and reassurance are very important. Many snakebites release only small, non-lethal doses of venom. The important thing after giving swift first aid is to *get help as quickly as possible.* If possible, and if help may take time to appear, improvise a stretcher and start carefully carrying the victim either to the nearest point where motor transport can make the pickup, or to a suitable area (bush clearing, flat hill area) where a helicopter can take him aboard. If this is done, make sure that the messenger(s) going for help knows your route and objective.

Antivenene for most of the more common venomous snakes is now generally available. This is why it is very important, if the snake has been killed, that it be kept for identification. Antivenene is usually injected intravenously by a doctor. So long as the patient is still alive make every possible effort to get help, or get to help. *It is never too late to give antivenene.*

ARTIFICIAL RESPIRATION (mouth-to-mouth resuscitation) may be needed urgently if signs of difficulty in breathing show themselves, and must be kept up until help arrives. External heart massage may be necessary and helpful at the same time. If paralysis appears in breathing or swallowing, the victim should be lain on his side, head down and turned aside. If paralysis shows in the tongue, it should be pulled forward so that it causes no obstruction to the limited breathing.

SUMMARY. Keep calm. Apply tourniquet immediately. Reassure victim and keep him and the affected limb immobilised. Send for help urgently. Organise carrying of the victim to suitable area if possible and necessary. Be ready to apply artificial respiration and heart massage. Keep the victim under close observation.

SPIDERS. The red-back spider, which is also known as the black widow in the United States and the katipo in New Zealand, has the unpleasant habit of cannibalising its male after mating. The red-back is easy to recognise. It has a small globular body of dark brown or black with a vivid red stripe down the centre of the back. The length of the body is about 1,3 centimetres. It has eight legs like the

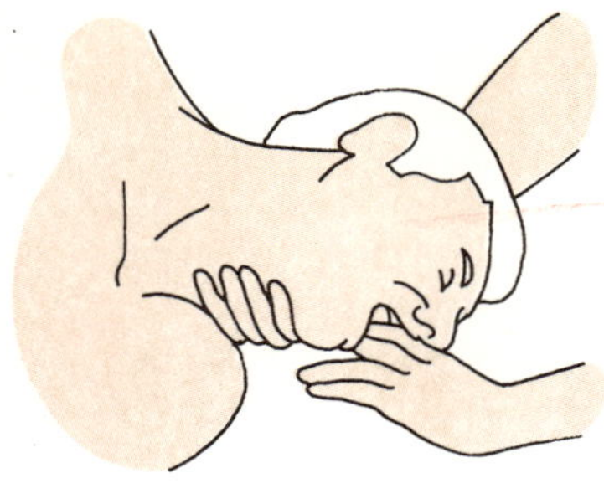

1. Check airway is clear behind tongue – no obstructions.

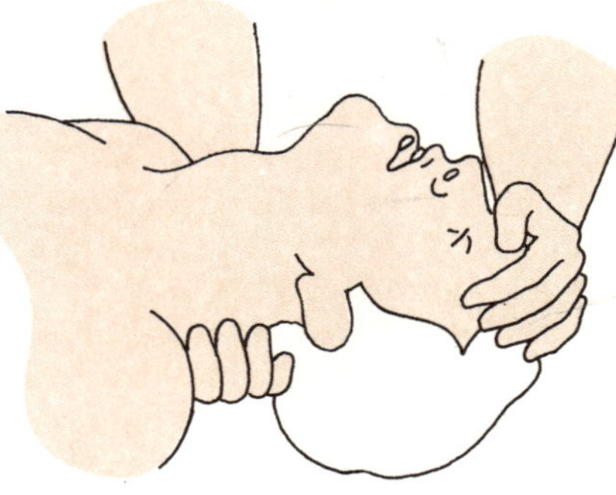

2. Supporting head, tilt it back so air can enter lungs.

3. Pinching patient's nose and still holding head back, blow strongly into patient's mouth.

4. Turn your head to see if patient's chest has risen, release nose and listen for any exhalation. Repeat until breathing resumes.

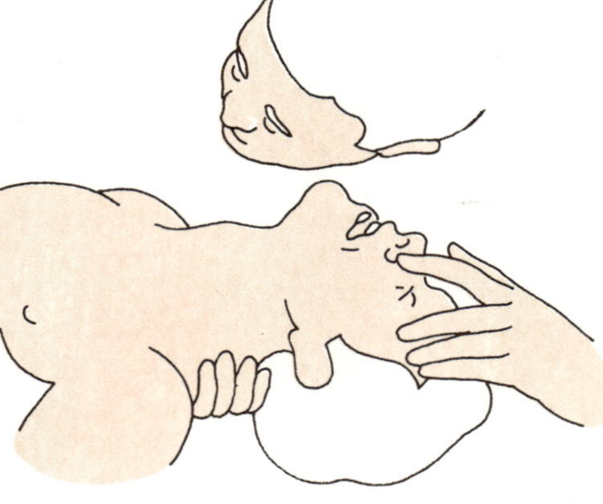

rest of the spider family and is a commonplace of backyards all over the continent, as well as elsewhere. The red-back favours shady situations protected from the elements, choosing rubbish heaps, flower pots, undisturbed boxes and tins in backyards, in wood heaps, under stones and bushes, beneath the loose bark of trees. The male is less than half the size of the female. Venom is injected into victims through channeled fangs. Despite its small size the venom it produces is strong and can have serious effects. A red-back bite can kill a child in a matter of hours and make an adult seriously ill for a week. The venom is strongest during the breeding season.

The bite of the red-back may or may not be felt. Usually the bitten area swells up and becomes red and inflamed. The burning pain which follows becomes intense and begins to spread. The victim experiences both sweating and shivering. Treatment is by anti-venene, but most cases recover, in time, if this treatment is not available.

The funnel-web spider has a large, long body and short legs, makes his home in a horizontal funnel-shaped web amongst loose leaves, roots or rocks, and has a bite that has, on occasion, proved deadly to adults. These creatures are most common in the sandstone areas around Sydney and occur further to the north. Similar spiders encountered elsewhere are believed to be members of the same family and are probably just as dangerous. The funnel-webs prefer a cool, dampish climate and may be found in forests which receive moderate rainfall. They cannot stand heat, dryness and sunlight, and usually hunt at night.

There is now an antivenene for the venom injected by the fangs of the funnel-web. The pain following a bite is extreme and quickly spreads to the rest of the body accompanied by sweating, delirium, and possibly respiratory failure. Treatment is the same as for snakebite. If possible a tourniquet is applied. Venom is removed from the area by sucking or washing. Medical help should be sought urgently.

SCORPIONS in Australia can be extremely unpleasant but are not lethal. The larger and more dangerous scorpions live well to the north of the continent. In the southern areas scorpions are small but widely distributed. The venom from a scorpion sting is neurotoxic like that of some snakes. A sting is extremely painful and may lead to a feeling of numbness which spreads to other parts of the body and causes the complete collapse of the patient. Treatment is with pain-killer tablets, hot compresses to the bitten area, and complete rest.

TICKS are small oval-shaped bloodsucking creatures about 5 milli-metres long which attach themselves to the skin and burrow their heads into the flesh to draw blood, on which they feed. The initial bite is hardly ever felt, but numbness and a swelling may develop. Continued attachment of the tick may lead to a general poisoning and ultimate paralysis. As well, ticks transmit very dangerous viruses and infections to human beings. If the tick's head is buried firmly in the flesh it is best not to try to remove the creature by pulling it away. This will leave the head still buried, a source of continuing complication and infection. The tick is best left in place until it can be soaked in turpentine, chloroform or other volatile liquid, when it will withdraw its head spontaneously and the whole creature can be removed, after which recovery should gradually take place.

Bush Tick

CENTIPEDES deal with their prey by injecting it with venom. They are harmless creatures, a few of which, if large enough, might inflict a temporarily painful bite. Centipede venom is not strongly poisonous, and has never been known to be lethal in Australia. Rest and hot compresses will reduce pain and promote speedy recovery. Few centipedes cause pain greater than that from a wasp or ant sting.

Red Back spider

MARCH FLIES, MOSQUITOES AND RED BULLANTS can all inflict bites of varying degrees of painfulness and subsequent irritation. None is dangerous but all cause acute discomfort and annoyance. There is now a number of proprietary brands of personal insect repellent on the market and a supply should always be carried into the outdoors if only for the sake of comfort.

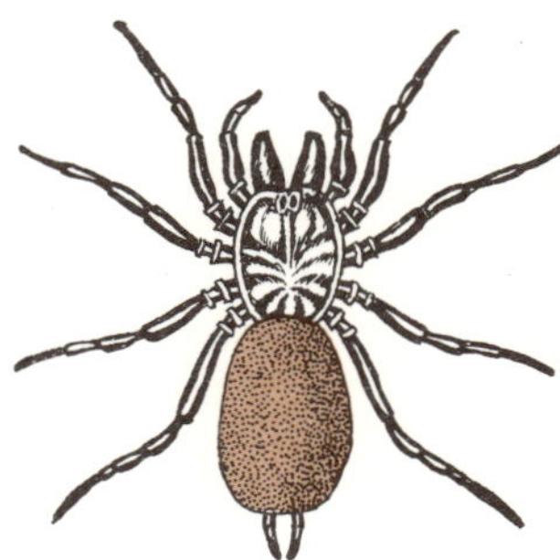

Funnel Web spider

LEECHES are slug-like creatures, usually encountered in damp areas that suck blood from the hosts (including human

Scorpion

beings) to whom they attach themselves. They can vary in colour between black, blue, yellow, and green, and in length between 5 millimetres and 15 centimetres. Leeches can circumvent most defensive arrange-ments of socks, boots and clothing to attach themselves silently and painlessly to their hosts. They cut open the skin and start sucking the blood while injecting a convenient chemical which ensures that the blood will flow freely and prevent coagulation. Usually, the first indication of a leech is a powerful itching sensation where the blood is being drawn. The leech may drink up to ten times its own weight of blood. They can live and grow on two meals, or even one meal of blood each year. The leech should not be pulled away from the skin without first forcing the withdrawal of the head by the application to the body of salt, or a heat source such as a glowing cigarette or a match end.

Proprietary insect repellents help in keeping off leeches, as do domestic soap and eucalyptus oil. A suggested repellent, if nothing else is available, is to rub the legs with young gum leaves. Soaking clothes and socks in salt solution has also been suggested as a solution to the problem.

F IRE is the greatest single danger of the Australian bush, and our first responsibility is to prevent bushfires starting by taking suitable precautions. Any cooking fire is best laid on rock and carefully sited so that sparks cannot reach any inflammable material (this includes trees, bushes, fallen leaves and debris of the surrounding bush itself) that may be nearby. If there is no rocky area then the site of the fire must be cleared of inflammable material for some distance all round. Stones, sand or earth should be arranged as a barrier around the fire, which should *never* be left burning without someone close by keeping a sharp eye on it. Always extinguish the fire completely after use. Use water, sand or earth.

In hot weather and very dry bush, smokers should stop smoking altogether or carry a small tin into which they can tap ash and stub out cigarette ends to be carried safely away. Habit is strong and pipe smokers especially are liable to throw down burnt match ends which may still be glowing. Pipes too should be emptied into a tin for safe disposal later. All smokers have a special responsibility either to take stringent precautions against starting fires or abstain altogether while there is a substantial fire risk. Campers and walkers travelling through the bush are subject to the same penalties as anyone else for lighting fires during a time of total fire ban. The penalties are very severe and include heavy fines and long periods of imprisonment. It is important therefore to carry at least some food which may be eaten without cooking. This is not a great hardship on the very hot days of total fire ban conditions, or when the bush is hot and dry enough to make fire lighting risky.

BUSHFIRES may happen anywhere in the bush and at any time in hot weather. The smoke of a bushfire is usually visible from far away and it is only commonsense to keep well away from it and from thick bush where fires may easily break out. Bushfire sparks can be blown for great distances unextinguished. A change of wind can set new fires going almost anywhere in the area. Think carefully before moving into country where bushfires are burning. Tomorrow is another, and probably safer, day. If the small beginnings of a fire are found in the bush it is obviously our responsibility to put it out as quickly as possible, thus preventing the loss of valuable forest, innumerable bush creatures, and even perhaps saving domestic livestock and human lives. If the fire is well under way it is futile to attempt to control it without expert knowledge and suitable equipment. Action to ensure survival is the first consideration: get away to safety at once.

Surviving bushfires, like other bush emergency situations, is first of all a psychological matter. Fire is the most terrifying of natural phenomena. Dense smoke, the roaring of the fire, the crackling of branches, and the intense heat borne on the wind, are bound to produce the beginnings of feelings of panic. These have to be firmly resisted. Panic is a prodigious waster of vital energy and leads straight to hysterical irrationality during which wrong, and perhaps fatal, decisions are bound to be made.

Try, as coolly as you can, to assess the situation. Is escape a possibility, or are you encircled by the fire? If it seems possible that you can get away move deliberately and carefully to get out of danger round the edge of the fire and behind it. Don't run unless it is absolutely essential. Try to hold your breath in thick smoke and fill lungs only in clearer, purer air. Cloth of any sort (preferably moistened — even spittle will do) held over the mouth and nose will mitigate the effects of smoke to some extent.

Try to pick a clean path unobstructed by fallen trees or other debris. If it is necessary to make a dash through flames, cover up as much as possible, inhale plenty of air, and go determinedly and without hesitation at the flames and through to the burnt area behind. If the flames are more than about 10 metres in depth, don't attempt to get through. If the flames are more than 2 metres in height, don't attempt to get through. If you decide after all to run away before the advancing flames, run downhill since fire moves more swiftly uphill. Again try to outflank the fire and get on to the burnt area behind it.

If encircled or trapped before an advancing wall of flame, light a backburn of at least 7 metres in length and move on to the ground it has burned over. If you are carrying an aluminium foil safety bag, get into it. While the foil cannot resist flames it will effectively reduce the radiated heat all around you. If you have no bag, cover yourself as best you can to reduce radiated heat. Ponds and running water are usually safe to get into but standing water tanks get hot quickly and collect fumes. A few minutes in such a tank will cause total collapse if asphyxiation by the accumulated fumes has not already occurred.

Without pond or stream, lie down and cover up as completely as you can. Get into any small depression that will hold you, or behind a log or rock where you can display the smallest possible surface area. Clear away any inflammable material beforehand. If it is possible bury yourself in loose soil or sand. Try to keep calm. Conserve your energy. Protect yourself against radiated heat. Panic is your greatest danger.

FIRST AID means exactly and literally what it says: in the case of serious injury it is initial temporary help that is given as an interim measure until skilled medical aid is available. Fortunately, serious accidents are not common, and basic knowledge of first aid is always useful in dealing with any minor happenings. Everyone venturing into the bush should carry a small personal first aid kit. Parties should carry in addition a larger and more comprehensive first aid outfit. Everyone venturing into the outdoors should have a copy of, and be familiar with the contents of a good first aid manual.

BLISTERS begin as small sorenesses frequently caused by badly fitting boots. Their pain and discomfort can be minimised if they are caught in this early stage, when a broad strip of adhesive tape should be smoothed over the spot and to cover a large area around it. This will ease the soreness and prevent a blister developing. Adhesive taping has to be done *at once.* Regardless of what anyone else may think, *stop* and see to it. A few minutes delay here may save hours and perhaps days of pain during which other people in the party are seriously inconvenienced.

If a blister does appear, cover with plaster and try not to interfere with it. Pricking with a needle to release fluid seems easy and helpful, but is tricky and likely to cause serious infection. If the blister has burst anyway, apply antiseptic cream, cover and protect with sterile bandage or dressing. Prevent blisters by wearing boots that have been properly broken-in to your feet, and never accept a loan of someone else's boots — only a sadistic and subtle enemy would offer them!

BURNS can be infected very easily. Small burns are best covered with a dry dressing and then left severely alone. Large burns should be covered with several layers of cloth (strips of lint or whatever is available). Avoid touching the burn area with the hands. Don't put grease or oil of any sort on the burn, and don't attempt to burst any blisters that may have formed.

SUNBURN occurs easily and is best prevented by use of a proprietary sunburn lotion. If sunburn still occurs treatment is with the sunburn lotion and by covering the area to prevent further exposure. If no lotion is available and the burns are extensive and swollen, compresses of cold tea and tea leaves may be applied. Failing this apply plain cold compresses. Sunburn can cause serious illness. It is a result of neglect and is easily prevented.

SPRAINS usually happen to ankles. If the sprain is severe with much local swelling, *leave the boot on.* Once off, walking becomes difficult or impossible and the boot cannot be replaced. With the boot on, the patient can walk, without too much risk of further damage, for considerable distances. For the first day after a sprain, keep applying cold water and cold compresses or both. After that apply warm and then hot compresses.

SMALL CUTS AND ABRASIONS are easily treated with antiseptic cream and Band-Aid type dressings after dirt has been removed and the area cleansed. It is important not to neglect small injuries like this because of the increased risk of infection. A quite small but badly infected wound can quickly become incapacitating if not treated properly.

LARGER INJURIES demand determined calmness and the application of basic first aid principles, the first of which is: never attempt to move an accident victim if you can avoid it. Make sure the patient can breathe and is breathing and listen for the heartbeat. Look for bleeding. If it is obvious and profuse don't be intimidated by it. Blood always looks worse than it is and quite small wounds can produce very large quantities of blood. Remember, movement may cause worse damage than is already present. Clean away the blood, locate the wound, and stop the bleeding by applying a dressing or pad of other material. If blood is spurting regularly from the wound and is bright red in colour (arterial bleeding) a dangerous amount of blood may be lost very quickly. Apply pressure to the wound with the bare hands if necessary while a tourniquet is applied. You *must* loosen the tourniquet for about half a minute in every twenty minutes, but a tourniquet should only be applied in cases of extreme necessity. Wounds should be cleaned with an antiseptic solution but the first necessity is to stop the bleeding. The edges of very large cuts may be drawn together and kept in place with sticking plaster which is applied in strips at right angles to the wound.

FRACTURES are extremely serious injuries. Compound fractures, in which the broken bone is protruding through the skin, need special care and urgently require skilled medical attention. If this is not forthcoming the limb has to be very gently straightened so that the bone slips back under the skin. *Do not touch the exposed bone in any way, and undertake this only if medical aid is very far distant or inaccessible.* If the bone end has picked up dirt, rinse it off with a solution of salt and clean water (a teaspoon of salt to four cups). *Do not touch.*

In all cases of fracture the limb must be immobilised by splinting in order to prevent further damage by movement. Arms should be splinted and then supported and fastened against the body. If no splint can be improvised for a broken leg it can be partially immobilised by tying it to the other leg.

SHOCK in most injuries is invariably present, or develops later. It should always be treated, as part of the treatment of the injury, by keeping the patient warm and at rest (preferably lying flat). Give warm drinks but no alcohol. Try not to let sweating occur; shock is made worse by sweating. Shock may be delayed. Look out for it as weakness, nausea, shallow breathing, weak pulse, cold clammy skin, blank eyes with enlarged pupils. *Always treat for shock,* as well as for the injury, immediately.

EXPOSURE (HYPOTHERMIA), FROSTBITE, HEAT EXHAUSTION AND SUNSTROKE are all dealt with in Chapter 12.

CAUTION. Where serious injuries have occurred, first aid aims only to keep the patient comfortable and alive until proper medical attention can be summoned or the patient moved to where it can be obtained. It is important to decide early whether the better course is to bring in a doctor or whether the condition of the patient and the resources of the party make it possible for a safe carry out.

The contents of this short section on first aid are no substitute for a training course with either the Red Cross or the St John Ambulance Association, just as a first aid course is not intended as a substitute for skilled attention by a doctor.

A PERSONAL SURVIVAL KIT should be carried by anyone going into the bush or into remote, difficult country. The kit must be small enough to fit into a pocket or pockets. It should include a whistle, preferably with a lanyard of some sort (leather, nylon) so that it can be attached firmly to belt or buttonhole or other anchorage. The lanyard is then available in extreme necessity as an emergency tourniquet. Matches, the cheap, readily available, waterproof sort, in their original box with striking surface, are always likely to be useful. A miniature compass of the cheapest simplest sort, so long as it is accurate, will not be more than 2.5 centimetres (1 inch) in diameter and weigh hardly anything. This is in addition to the regular large Silva-type compass which should always be carried anyway but may get lost, damaged, or broken. Half a dozen Band-Aid strips can go into the odd corner of almost any pocket with a few instant energy glucose tablets. An old diary, the sort that includes a pencil in the spine, can be used to leave messages, to record land-marks, and at a pinch a page or two may make all the difference between being able to light a fire and not being able to light a fire. A good knife, which can be an emergency kit in itself, should be carried. The best knives that come with a variety of options in the tools they carry, are the original Swiss Army knives. They are strong, light, efficient, but rather expensive.

A knife is only as good as the pocket sharpening stone that goes with it, and a small sharpening stone should go with every good knife; not necessarily as part of the personal survival kit. The knife can be easily kept very sharp with a few strokes on the stone. The best way is to have the back of the blade about 3 millimetres above the stone when the cutting edge is touching it. Holding this fractional angle, move the blade forward as if cutting into the stone. Turn the blade over, maintaining the same slight angle after each stroke.

A final but important inclusion in the personal survival kit is a water purification kit which is available cheaply from chemists and consists of a supply of two sorts of small tablets. The first tablet is put into the water bottle and left for thirty minutes after which the second tablet, of a different colour, which takes away the taste of the first, purifying tablet, is inserted. The resulting water is safe to drink and without any detectable chemical taste.

A few hooks (size 9 or thereabouts) and a 10 to 15 metre length of nylon of 6 kilograms breaking strain, complete the personal safety kit, all of which will go into a flat tobacco tin which can be slipped easily into a pocket.

Further Reading

D. Underhill, *Australia's Dangerous Creatures*, Reader's Digest, 1987.
E. Stodart, *Biters and Stingers*, Octopus, 1989.
Bay Books, *Dangerous Australians*, 1985.
H. Frauca, *What Animal is That?* Doubleday, 1982.
S. K. Sutherland, *Australian Animal Toxins*, Oxford University Press, 1983.
St John Ambulance Brigade: *First Aid Manual,* latest edition.

CHAPTER 16

THE WHITE WORLD: BASIC CROSS-COUNTRY SKIING

When the snow lay round about
Deep and crisp and even.
CAROL

What was probably the first ski club in the world was formed at Kiandra 1650 metres up in Australia's Snowy Mountains in the early 1860s. It was called the Kiandra Snow-Shoe Club but the 'snow-shoes' were in fact crude skis made from sawn timber with a number of grooves underneath. Club members wore no special skiing clothes but skied around in suits with watch-chained waistcoats, stiff collars and bowler hats. Gold brought the skiers to Kiandra. When the gold ran out, as it soon did, skiing remained, and has led over the years to the present alpine villages of the Snowy Mountains and the Victorian alps with their tows and chairlifts and downhill skiing.

The mountain snowfields of Australia are larger than those of Switzerland and are, as yet, hardly touched. They are also, because of the high plateau nature of much of the country, more suited overall to ski touring or cross country skiing than to downhill skiing. With the opening up of the ski resorts there has been a growth of complications in downhill skiing equipment which has added greatly to the expense. The resorts are crowded each winter. The adjacent downhill ski-runs are limited. Long queues form for the lifts and tows, and fashions in expensive clothing seem to have become the preoccupation of the majority of the people who stay at the resorts.

In recent years there has begun something of a revolt against downhill skiing and its noisy social associations. The people who wanted to ski and experience the beauty of the snow-hushed high country, in preference to preening themselves in queues for much of their days, took up cross-country skiing. Equipment for cross-country skiing is cheap compared with that for downhill skiing, the skiing itself is easier, clothing is utilitarian, and there are no social obligations.

Snow is a miracle. Out of all the countless billions of snowflakes that fall no two are of the same pattern! A familiar landscape revealed suddenly in stark black-and-white is not diminished but takes on a new dimension. If one is properly clothed and equipped there is an immense exhilaration in the crisp, biting cold air and the exploration possibilities in the dramatically changed mountain

world. To really know the winter hills, to be at home in the austere splendour of the snowscape one must be a cross-country skier.

THE DANGERS of this sort of skiing should not be minimised. Blizzards and bad weather are always a threat at high altitudes in winter. Snow and mist can reduce visibility to nil in a few minutes. Accidents — sprain or fall or fracture — can happen in areas remote from help. Cross-country skiing is not for solitary skiers but for strong parties which (if any real distance is to be covered) must be made up of a minimum of three people. First-class navigation is essential in the very different looking snow world. Basic competence in skiing is the foundation on which safe cross-country skiing is built. A nose for weather changes, and a snowcraft that can quickly construct a shelter in time of trouble and deal with difficulties, are desirable attributes in members of the party.

THE EQUIPMENT for cross-country skiing is very different from the massive, rigid 'spaceman' boots, large bindings, and broadbased skis of the downhill skier. Cross-country skis are light and narrow, boots or shoes are lightweight and are in the binding only at the toes so that the heel may lift and the ankle may have movement.

There are three sorts of cross-country skis available. The *touring ski* which is rather wider, stiffer, tougher, and heavier but slightly more stable than the others. The *light touring ski* is a narrower, lighter version of the touring ski and is often recommended for learners. The *racing ski is* a highly specialised super-lightweight touring ski that has been developed for racing on prepared tracks. Bindings are at the toe and made of metal, with plates for the heel. Boots are lightweight and very flexible. Some are much more like shoes than boots. Stocks or poles need a little whip in them to help forward propulsion. They are made from bamboo, metal or fibreglass. The correct size for skis is found by raising the arm to a vertical position when the ski should just reach the heel of the hand. Stocks are measured by tucking the ends under the armpits when the *baskets* should be resting on the snow.

WAXING is part of the preparation of the skis. There are many waxes, each of which suits a different sort of snow. Most skis now available do not need wax. These have soles of sealskin or other pattern and are said not to be as good as well-waxed skis. The purpose of the waxing and the patterns is to enable the ski to safely grip the snow when stationary or climbing hills, and to slide easily forward when moved. Some beginners prefer waxless skis because they enable them to get valuable experience without too much bother. Later, as their skill increases, they change to waxed skis.

The correct waxing of skis, to match them to current snow conditions, is a skill that comes only with experiment and practice. It is both an art and a craft, and exercises a strange fascination as well as providing the matter for endless discussion and controversy. Waxing begins with the sealing of the wooden ski bottoms with a pine-tar base. There is a variety of base compounds available which can be applied by hand or be burnt on with a small blowlamp. The aim is to get the wood to soak up as much of the compound as possible so that there is effective adhesion to the ski and a sound base for the running waxes. Skis of synthetic material will not of course absorb base tar compounds unless specifically stated to do so.

Waxing requires some warmth. The running wax is dabbed on all over the surface and smoothed to a fine finish with a cork block or spreader. A number of thin layers is applied. All the running surface has to be waxed. The correct match of wax and snow is the secret of good waxing and there has built up around the waxing process a craft-mystique which involves innumerable sorts of waxes for every imaginable state of snow. Ancillary tools of the waxer's trade include torches heated by petrol, butane, and propane gas, waxing corks, scraper-corks, scrapers, waxing irons, and thermo-meters, as well as complex charts indicating the best waxes for the most common sorts of snow (dry powdery, granular, refrozen, mushy and so on).

After learning from a teacher with years of practice behind him, the best way of learning waxing is by doing it yourself. Mistakes are, as always, the best part of a waxing education. Most success is founded on frequent failures and ways developed to overcome them. Theory is more hindrance than help until experience has been acquired practically. The difference between well and badly waxed skis is astonishing and is the best of all lessons to learn from.

WAXLESS SKIS are used almost universally amongst cross-country skiers. The ski bottoms are moulded into step, fish scale, or similar patterns, or have strips of mohair fixed to them, which function like wax. Waxing is confined almost completely to ski racing enthusiasts.

CROSS COUNTRY SKI CLOTHING differs from winter walking clothing only in the use of knee breeches and long socks, for the comfort and freedom of movement they give. An outer hooded parka (and trousers) and the usual layers of light clothing are sufficient for most conditions.

In cross-country skiing, layers of light clothing, which can be quickly donned or shed, are important in avoiding both sweating and getting chilled.

METHOD. Cross-country skiing is based on walking. It begins with shuffling the skis past one another until confidence builds up in the efficacy of the grip of the ski on the snow when the initial shuffle develops into a limited push-and-glide, push-and-glide, which leads to knee bending, a leaning of the body forward, and a loping step which covers the ground much more quickly than walking. This, in essence, is what cross-country skiing amounts to. The rest is practice, concentration, and the development of the highly polished technique of the racers, which is a specialisation unnecessary for ski touring.

FIRST EXCURSIONS, after confidence and sound basic technique have been achieved, are best undertaken for a half-day or a day, in company with an experienced cross-country skier who can supervise the novice's efforts and draw attention to difficulties and possible dangers. Minimum party size is three for safety, and spare ski-tips and fixings should be carried. Ski-tips can break as easily close to base as anywhere. Sunglasses are an essential always, and gaiters are useful in preventing snow getting down between socks and boots if the snow is deep. It is important to drink frequently, since the rate of dehydration, from effort and from dry air absorbing moisture in the lungs, is considerable. Food is a matter of personal preference but nibbles and quick energy tablets should be carried in addition to

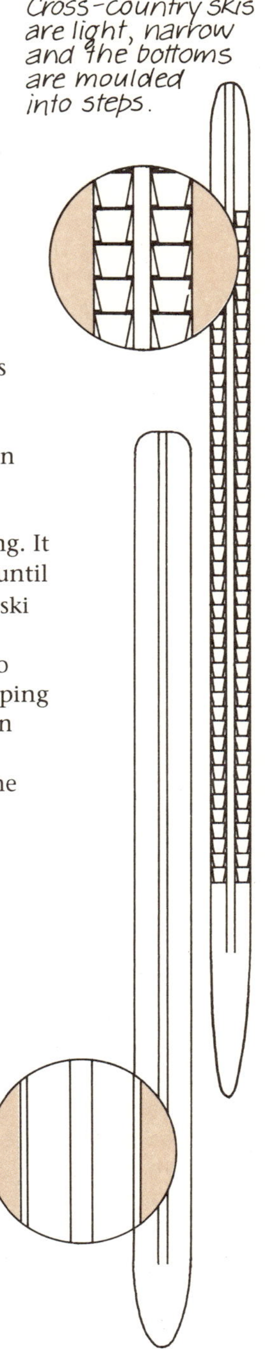

regular rations both for the physical boost they give and for the enjoyment.

LONGER EXCURSIONS become serious expeditions, demanding proper preparation and first-class equipment. There are numbers of skiers' huts and cattlemen's huts dotted around the snow country which provide primitive accommodation and which can save the carrying of tent, groundsheet and pegs. Sleeping bags should be of the best, box-quilted down, with a strong light outside cover for extra warmth. Amenities in the huts are very limited and it is best to carry either a foam-plastic hip pad or half-size inflatable air bed with a lightweight plastic sheet for use as a ground sheet.

CAMPING IN SNOW demands experience and toughness. A tent with a sewn-in groundsheet and adequate closure against the finest snow is necessary. It is better not to cook inside the tent if possible. Liquid fuel stoves are usually preferred to gas in these conditions since gas cartridges tend to ice-up in extreme cold. Adequate food is very important. Hot food boosts morale more than cold food but delivers up much the same amount of warmth and energy. Newcomers to snow camping should *always* make their first expeditions under the leadership of an old hand at the game. To do otherwise is to invite serious trouble with possible fatal consequences.

IN AN EMERGENCY it may be necessary to make a temporary shelter. In order to save precious energy look for small areas already protected in the lee of trees, low bush or large boulders. Often snow can be removed to make a hole in which to squat, which can be covered over with branches or anything else that is available. Alternatively, on slopes, a snow cave can be dug into deep snow. This requires much more energy and care, and it is important to make sure of being able to dig yourself out if the entrance collapses or becomes snowed-in. A good cave consists of a low narrow tunnel, several metres long, leading into a larger chamber in which there is room to rest comfortably. Adequate ventilation is essential. It helps to have the tunnel entrance slightly lower than the chamber so that warmed air (which tends to rise) is not being steadily drained away. In sufficiently deep snow a foxhole may be dug with a narrow entrance and room to stretch out lower down. A piece of plastic can be used to close the entrance so long as adequate ventilation is ensured. The lower down the mountain the shelter can be constructed the better. Avoid high exposed places if possible.

Once in shelter, concentrate on keeping warm by insulating your body from the ground with anything that can be spared from the rucksack. As long as clothing is in direct contact with snow or ice heat is being conducted away from the body. Put your feet in the

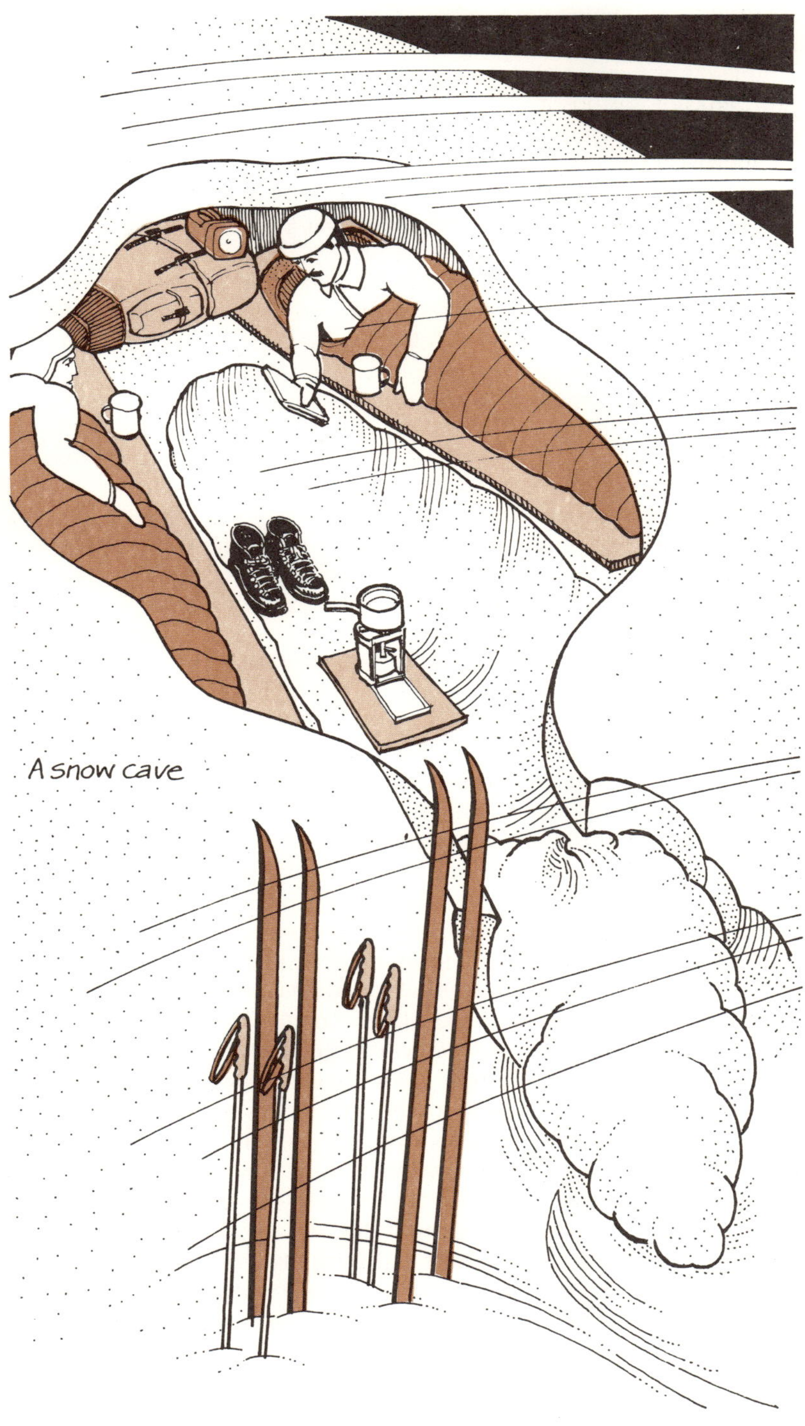

A snow cave

rucksack, and if you have a foil or plastic emergency bag, get into it and arrange the top so that it covers your head with just a small airhole opposite your mouth. Keep huddled together. Eat, and have hot drinks (if it is possible to make them) at regular intervals. Any cooking will use up the air in the shelter and extra ventilation should be improvised before the cooking starts.

Making a shelter in snow is an arduous task even with the proper lightweight tools. Without them, using improvised implements, it is a major work demanding large amounts of effort and energy which might be better used in keeping members of the party warm. To get an idea of what is involved it is worth spending a couple of hours on a fine touring day to experiment in shelter making, and then try to imagine the extra difficulties and effort involved for a tired party in blizzard conditions.

FOOD for snow travelling is the chief source of heat for the body. Large regular meals of easily digested food are the best heat producers. These should be supplemented with plenty of quick energy nibbles like chocolate, raisins, boiled sweets, glucose tablets and sweet biscuits. Carbohydrates are the main providers of energy. Fats also contain much more energy than most other foods, so it is important to see that a snow diet is rich in fats as well as sugars, without omitting the other foods that help to make up a balanced diet.

CROSS-COUNTRY SKI AREAS in Australia extend from the Victorian alps to the northern end of the Snowy Mountains near Canberra. The Victorian snowfields include the Mount Baw Baw area, Stirling, Mount Buller, Lake Mountain, Mount Hotham, Mount Bogong and Mount Buffalo areas. There are large ski villages at Mount Buller and Falls Creek (Mount Bogong area).

In New South Wales, the snowfields are generally in the Kosciusko National Park from the country round Kiandra on either side of the range southwards to the Victorian border with ski villages at Thredbo and Perisher and a very large chalet at Charlotte Pass.

Snow in the Victorian alps lasts from three to five months, the season usually beginning in June. Snow extends upwards from 1000 to 1200 metres at a depth of 30 to 100 centimetres with drifts several metres thick. Conditions vary from year to year and when a severe winter is experienced snow may begin as early as April and extend far below the usual snowline. Fog is common over snow in winter and leads to 'white-out' when ground, sky and air all look alike and acute feelings of disorientation are experienced. Winds of 100 kilometres an hour can occur with very much stronger gusts. Wind and snow or wind and rain make the most unpleasant and difficult conditions.

In the Snowy Mountains in New South Wales snow falls

occasionally in all months of the year but falls regularly and substantially for the six months from May to October. Most substantial falls are from 1200 metres upwards but the snowline can begin much lower down. Winds of up to 166 kilometres an hour have been recorded in the area. Drifts up to 5 metres have been measured, but very much deeper drifts are known to exist in favoured but inaccessible places. As in the slightly lower Victorian alps, which are a continuation of the Great Dividing Range to the south-west, a particularly cold wind may extend the snow season for the eight months from May to December.

Australian snow is moist and soft more often than not. This is because temperatures in the daytime usually rise a little above freezing. Very low night temperatures can turn the melt into a forbidding sheet of ice. This general wetness of the snow is worth keeping in mind when purchasing ski waxes. Obviously larger amounts of wax suitable for this condition will be needed than for others.

So long as there is plenty of snow, Mount Buffalo, which is a separate high plateau north of the Victorian alps, is a particularly good beginners' ground which is being developed for ski touring by skiers from nearby Wangaratta, Albury and Wodonga. Trips over the plateau can be varied from short and very easy ones to long and quite tiringly difficult ones. The area is a particularly safe one because of the limited size of the plateau but still demands *all* the normal equipment, skills and precautions.

THE RISK OF HYPOTHERMIA is always greater in snow country than elsewhere. On even the shortest trips, wind- and waterproof parka and overtrousers should be carried, together with spare warm woollen clothing, personal survival and first aid kits, and food and nibbles. Wet clothing is the great enemy. Cold and wind working on the wetness can take away great amounts of body warmth very quickly. In snow a constant lookout must be kept for the signs of hypothermia in all members of the party.

Further Reading

Henry Plociennick, *Australia's Snowfields*, Hamlyn, 1976.
W. Cross, *Australian Skiing*, Roscope Publications, 1983.
I. Hampel, *Australian Cross Country Skiing*, Kangaroo Press, 1988.
Sports Illustrated Cross Country Skiing, H & R, 1984.
N. Gillete, *Cross Country Skiing*, Diadem Books, 1985.

WEATHER

There is really no such thing as bad weather,
only different kinds of good weather.
Lord Avebury

WEATHER FORECASTING is a fairly inexact sort of science even today, when satellite pictures of cloud systems and the movements of air and moisture accompanying them are available for most of the earth's surface, and the nightly television weather report shows their positions. The weather maps we see in the media are more useful really for getting a rough general idea of how the weather is likely to behave in the next day or two, than for forecasting, with reasonable accuracy, the local weather during the next twelve hours. General weather forecasting has improved immeasurably in recent years. Local forecasting is still in the province of the local person interpreting the weather signs that people have noted since time immemorial. Nevertheless, the modern weather map is immensely helpful to outdoors people. With the most recent weather pattern filed in the back of the mind it is much easier to decide, for instance, whether the light rain that has begun falling is a local shower or the beginning of the passage of a cold front with accompanying continuous widespread rain that may make going difficult for the next twenty-four hours or so.

All life depends on the terribly thin skin of atmosphere surrounding the earth, shielding us from lethal cosmic rays, insulating us from the unimaginable cold and darkness of outer space, providing the air we breathe, distributing and redistributing the earth's water. The effectiveness of the atmosphere as a support and protection system for human beings is limited to a mere 12 000 metres. Higher than about 9000 metres, oxygen and special protection are required to maintain human life, because of the thinning of the air, the fall in temperature, and the decline in pressure. Climate and weather are born out of the lower levels of the atmosphere between ground level and about 18 000 metres. Above this layer is the stratosphere, a stable region which has little effect on the weather, and beyond that, thinning into the hostile emptiness of space at about 32 000 kilometres, is the still mysterious region known as the ionosphere.

CLIMATE is the pattern of what happens in the lowest level of the atmosphere over a wide area over a long period of time. Weather is what happens in the lowest level of the atmosphere, over a small area, in a short period of time. Weather is the concern of walkers, though occasionally a knowledge of the climate of the region is useful. The lower atmosphere is made up of masses of air of varying temperatures, different degrees of moisture, and the changes in pressure that result from their movements and interactions as the atmosphere circulates under the influence of the movement of the planet.

In the past two decades we have become increasingly aware of the threat to the thin skin of the earth's atmosphere from human activities. The portents are dire. The ozone layer is being steadily depleted. Huge holes in the ozone layer have opened up over Antarctica and the Arctic. The Antarctic hole has already begun to affect Tasmania and Southern Australia with levels of ultra violet radiation rising sharply over the past decade accompanied by a quite catastrophic rise in the incidence of skin cancers.

The major Australian cities are suffering from acute industrial pollution problems which are the cause of innumerable minor illnesses and are life threatening to asthmatics and people of all ages with lung problems. Poor farming techniques have led to widespread land degradation and salinity. Logging and wood chipping have raided our precious forests beyond even the minimum we shall require in the future. Disposal of waste and sewage has polluted huge areas of coastline and toxic industrial wastes are ruining creeks and rivers while refuse dumps are releasing increasing amounts of methane gas into the atmosphere and worsening the depletion of the ozone layer.

Elsewhere throughout the world, but especially in the developed countries, pollution has become a major problem with widespread acid rain ruining vast tracts of forest and reducing lakes to pale empty lifelessness. In Africa and in the Amazon basin of South America the tropical forests, almost literally the life preserving lungs of the earth, are being burnt out and destroyed at frightening speed.

Inevitably all this must, in the long run, affect climate and weather. There are suggestions that climatic change is already under way, that the planet is warming slightly. In recent years there have been a variety of anomalous weather patterns: aberrantly warm winters and recurrent hurricanes in Northern Europe, droughts in the USA and Africa.

The experts are agreed that *something* is happening. Both global warming and the beginning of a new ice age have been forecast. It seems impossible to say what exactly is likely to occur in the future. All that can be said is that both climate and weather are showing radical signs of change, and that seasons that could once be generally relied upon to recur with dependable frequency can no longer be relied upon to do so. In planning ahead for outdoor trips it is worth keeping all this well in mind.

AUSTRALIA'S WEATHER comes from the west. The system of high and low pressure and the fronts between them move in from the Indian Ocean region. As they move they are monitored and reported on by ships at sea, aircraft, weather stations, and orbiting satellites. The information provided is put together by the meteorological offices and made available as weather forecasts with accompanying weather charts. This complicated operation takes time and the information has then to be made available to the media, which takes more time. It is important to remember this when consulting weather maps and to make an allowance of about twelve hours for the delay.

WEATHER MAPS show the weather approaching from the west (it is actually about twelve hours closer than can be shown) by lines (isobars) joining together points of equal atmosphere pressure. The closer the lines the greater the wind, and vice-versa. These resolve themselves into low pressure areas in which the wind blows clockwise, and high pressure areas where the wind blows in an anti-clockwise direction. On the whole the highs bring fine weather and the lows bring unsettled weather and a much greater chance of rain. As these masses of air come in over the Australian continent they are modified and influenced by many things, including the movements of temperature of the earth, the density or absence of vegetation, and

High and low pressure areas.

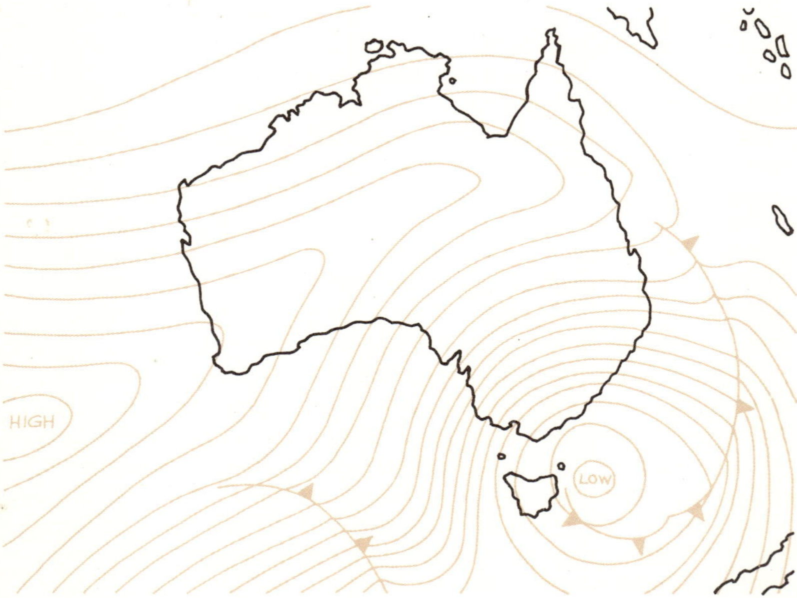

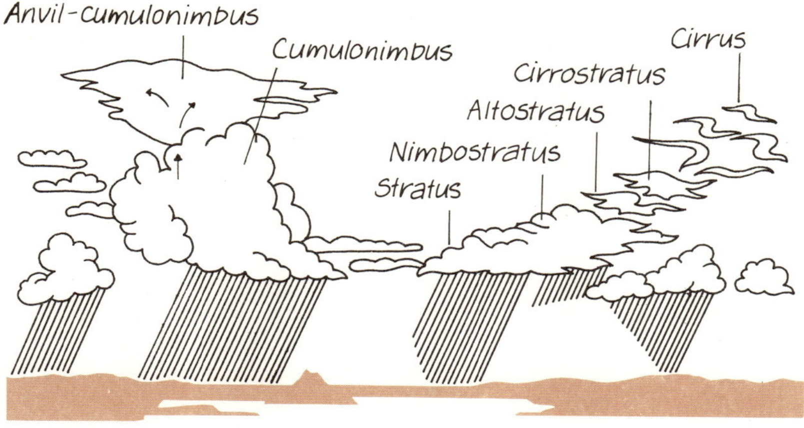

the presence of high mountains. The nature of the air mass, whether it is warm and humid, warm and dry, cold and moist, mild and dry, depends upon the area in which it forms. Air masses forming over central Australia are usually very hot and dry in summer and cold and dry in winter. Summer winds from the centre of the continent have a dry furnace feel to them and result in heatwaves.

LOW AND HIGH PRESSURE AREAS encounter one another in either warm or cold fronts. A cold front results in lowered temperatures and changed wind direction (counter-clockwise). As the air mass following the front moves in the barometric pressure starts to rise. Warm fronts passing through result in stabilisation of falling pressure, rising temperatures, and again counter-clockwise wind changes.

CLOUDS. High clouds, like the other clouds, can be valuable to the local forecaster. In Australia high clouds are those above about 6000 metres and are made up mostly of ice-crystals. The highest cloud at about 10 500 metres is called *cirrus* and is made up of fine filaments which are often called mares' tails. Lower down at about 7500 metres there occurs *cirrostratus* in the form of a veil of thin, almost imperceptible filaments covering much of the sky. In between *cirrus* and *cirrostratus* there occurs *cirrocumulus* known to most people as a mackerel sky.

Middle clouds are *altocumulus,* puffy rolls of cloud piled in bands at around 6000 metres and *altostratus,* a hardly varying blanket of darkish cloud at about 4000 metres.

Low clouds only extend upwards to about 2500 metres. *Stratus* is a uniform, low, grey cloud that makes what people call a leaden sky. *Nimbostratus* are rain clouds and are darker than *stratus*. *Stratocumulus* is thick, wet-looking cloud cover like dark cotton wool. It is lighter in colour when thin, and occasional patches of blue sky may be seen through it.

Vertical cloud development is exemplified in *cumulus* and *cumulonimbus* which can extend upwards through 5000 metres or more. *Cumulus* is familiar from warm summer conditions when warm moist air rises over the earth puffing the cloud upwards in fantastic everchanging shapes that are brilliantly snow-white on their summits. *Cumulonimbus* are what most people recognise as thunderheads. They develop from *cumulus* and extend themselves upwards to great heights, often in impressively dramatic shapes. Sometimes high winds in the upper air flatten out their tops, when they are called *anvil cumulonimbus*.

A BAROMETER, which indicates the rises and falls in atmospheric pressure, is almost indispensable to the local forecaster. A few people, for reasons not entirely understood, are sensitive to changes in barometric pressure which they experience as unpleasant physical sensations or as headaches, if a deep low is approaching, but the accuracy of a barometer is preferable to their, inevitably, subjective comments. Daily observation of the barometer readings and their relationship to the weather as it changes, and the clouds as they pass and alter their shapes, is the best preparation for local forecasting in the field.

LOCAL FORECASTING depends on observation of clouds and wind direction. If the movement of highs and lows from the west is slow, as it often is in summer, one can set out with a fair, general idea of the weather for several days. The movement of the highs and lows can be roughly monitored by remembering that in the southern hemisphere, with the wind at one's back, the high will be to the left and the low to the right.

BAD WEATHER is commonly associated with lows and usually announces itself by the changing of the clouds, which thicken high up, gradually overspread the sky, and form into dark grey, rather threatening woolly masses. Often *cirrus* changes to *cirrostratus*, then to *altostratus*, and finally to *nimbostratus*. A mackerel sky (*cirrocumulus*) usually indicates the approach of bad weather. The cloud movement is usually accompanied by a wind change. If a pocket barometer is available and is showing falling pressure rain is almost certain. Other bad weather indicators are: sharply rising winds, red morning sky, low-flying birds and insects, marked increase in cloud, darkening of the sky, a halo round the moon (rain), thunderheads *(anvil cumulonimbus)* building up, a rise in night temperature and a falling barometer.

GOOD WEATHER is commonly associated with highs, and is also heralded by rising barometric pressure, falling temperature, rising cloud, and clearing skies. *Cumulus,* in its small broken up form as well as in larger masses, suggests continuing fine weather. Other fine weather indicators are: steady barometric pressure, brilliant fiery sunsets, light steady winds (displaced in late morning on the coast in summer by an onshore breeze, often vigorous, which moderates towards evening), high flying birds and insects.

CONTINUOUS OBSERVATION is more important than large amounts of meteorological theory. People get caught out by bad weather because they forget that an eye has to be kept on the clouds, and the sky, and the wind. Nature seldom indulges in really rapid, dramatic changes. There are almost always clear and obvious indications of bad weather for at least several hours beforehand, and adequate time to turn back or make for cover. Only experience and daily observation, at home as well as in the field, of weather maps and forecasts as well as local weather indicators, can make you a really competent forecaster in the field.

The most useful single bit of information about weather is simply that when rain starts falling unaccompanied by wind, a long period of bad stormy weather with both heavy wind and rain will follow. If there is wind first and then rain the skies will soon clear to fine weather. This is vitally important to the bushwalker, who often has to decide at the onset of rain whether to press on, turn back, or make for cover.

REMEMBER. Mountains make their own weather and the resulting changes are quicker and usually more intense than anywhere else. Always keep a vigilant weather watch in highland regions.

Further Reading

S. Dunlop, *The Hamlyn Guide to Weather Forecasting,* Hamlyn, 1982.
Atkinson and Gadd, *Weather: A Modern Guide to Forecasting,* Mitchell Beazley, 1988.
F. Christie-Mitchell, *Practical Weather Forecasting,* Hutchinson, 1977.
Falk, *The Greenhouse Challenge,* Penguin Books, 1989.
Boyle and Ardill, *The Greenhouse Effect,* New English Library, 1989.
J. E. Lovelock, *Gaia,* Oxford University Press, 1979.
J. E. Lovelock, *The Ages of Gaia,* Oxford University Press, 1989.

INDEX